Applications of Field Theory Methods in
Statistical Physics of
Nonequilibrium Systems

Applications of Field Theory Methods in
Statistical Physics of Nonequilibrium Systems

Bohdan Lev
National Academy of Science of Ukraine, Ukraine

Anatoly Zagorodny
National Academy of Science of Ukraine, Ukraine

World Scientific

NEW JERSEY · LONDON · SINGAPORE · BEIJING · SHANGHAI · HONG KONG · TAIPEI · CHENNAI · TOKYO

Published by

World Scientific Publishing Co. Pte. Ltd.

5 Toh Tuck Link, Singapore 596224

USA office: 27 Warren Street, Suite 401-402, Hackensack, NJ 07601

UK office: 57 Shelton Street, Covent Garden, London WC2H 9HE

Library of Congress Control Number: 2021005226

British Library Cataloguing-in-Publication Data
A catalogue record for this book is available from the British Library.

ISBN 978-981-122-997-8 (hardcover)
ISBN 978-981-122-998-5 (ebook for institutions)
ISBN 978-981-122-999-2 (ebook for individuals)

For any available supplementary material, please visit
https://www.worldscientific.com/worldscibooks/10.1142/12091#t=suppl

Typeset by Stallion Press
Email: enquiries@stallionpress.com

Preface

In this book, we formulate a unified approach to the description of interacting many-particle systems combining the methods of statistical physics and quantum field theory. The strong point of this approach is that it may be applied to the description of phase transitions with the formation of new spatially inhomogeneous phases. Another advantage concerns the possibility to describe quasi-equilibrium systems with spatially inhomogeneous particle distributions and metastable states.

The validity of the methods of statistical description of many-particle systems and models (theory of phase transitions included) is discussed and compared. The idea of using the quantum field theory approach and related topics (path integration, saddle-point and stationary-phase methods, Hubbard–Stratonovich transformation, mean-field theory, and functional integrals) is also described in sufficient detail for understanding as well as for further applications. To some extent, the book could be treated as a brief encyclopedia of methods applicable to the statistical description of spatially inhomogeneous equilibrium and metastable particle distributions.

The additional strength of the book is that we not only formulate the general approach, but also apply it to solve various practical and important problems (gravitating gas, Coulomb-like systems, dusty plasmas, thermodynamics of cellular structures, nonuniform dynamics of gravitating systems, etc).

The book could be useful not only for students, but also for researchers dealing with various methods of equilibrium and nonequilibrium statistical mechanics.

Contents

Introduction

The statistical description of interacting many-particle systems remains one of the key issues of theoretical physics. Such description is needed for understanding the thermodynamic, kinetic, electronic, and electromagnetic properties of a wide class of new substances that have been intensively studied recently (liquid crystals, dusty plasmas, low-dimensional structures in solids etc). Many traditional problems, e.g., phase transitions accompanied by the formation of structures and metastable spatially inhomogeneous states, as well as the description of stationary states in open systems, also require further studies.

The main purpose of this book is to propose a new approach [1–6] to the statistical description of a system of interacting particles with regard for spatially inhomogeneous particle distributions. To describe such structures, it is necessary to work out a method that would enable us to select the states with thermodynamically stable particle distributions in the partition function. One such method makes use of the representation of the partition function in terms of a functional integral over auxiliary fields that makes it possible to employ the quantum field theory approach [7–14]. An attempt to apply the functional integral to the description of many-particles systems was discussed for the first time in [15]. Both advantages and challenges of this approach were described in [7]. Particularly, the advantage is that the extension of the functional integrals to the complex plane provides a possibility to apply the saddle-point method with no use of the perturbation theory. It allows to select and separate the system states associated both with homogeneous and inhomogeneous particle distributions.

The description of the formation of spatially inhomogeneous particle and field distributions is of great importance for condensed matter physics. It concerns both the physical understanding of optimum states of the system and is valuable for applications in practice [16–18]. Earlier investigations of the formation conditions and behavior of inhomogeneous states have mainly employed the statistical theory of nonequilibrium processes. However, spatially inhomogeneous particle and field distributions can also be formed in equilibrium systems. The conditions for the formation of such structures and their physical manifestation are determined first of all by the type of interaction. So, we have to formulate an adequate mathematical method that would describe the formation and behavior of spatially inhomogeneous equilibrium particle and field distributions.

A few model systems of interacting particles are known, for which the partition function can be evaluated exactly, at least in the thermodynamic limit. In this book, we demonstrate the efficiency of the proposed approach by a nonperturbed calculation of the partition function for the known model systems with interaction (hard spheres model, Coulomb gas, gravitating gas, etc.). This approach makes it possible to describe any system of interacting particles with regard to spatially inhomogeneous particle distributions. A typical physical situation that involves bound states in a particle system occurs when the interaction consists of long-range attraction and short-range repulsion. Another realistic situation is associated with the opposite case when the repulsion range is longer than the attraction range. Such physical systems are, e.g., electrons on the liquid helium surface [19]; polar atoms and molecules on a metal or dielectric surface [20, 21]; ions implanted in silicon [18, 22]. If interaction of this type occurs, the system cannot be homogeneous and hence it involves spatially inhomogeneous particle distributions, namely, finite-size clusters.

The proposed approach makes it possible to describe such particle distributions, to calculate cluster sizes, to estimate the number of particles within a cluster, and to find the temperature of the phase transition to the state under consideration. The number of particles in a cluster and the size of the latter depend on the interaction type and its intensity as well as on external parameters. The residual interaction (uncompensated after the cluster formation) produces interaction between clusters that, in turn, can cause formation of new spatial structures in the cluster system. The problem of how to find the cluster distribution and to estimate the influence of the external factors also can be solved in terms of the proposed approach.

Particularly, an approach to the statistical description of interacting particles and phase transitions accompanied by cluster formation was proposed in [3, 23]. Clusters are described by the function of spatial particle distribution. This function is a soliton solution of the nonlinear equation that arises in most cases of statistical description of interacting particles. The method developed in the papers [3, 23] is based on the application of the quantum field theory approach [8, 9, 13, 14, 24–26] and provides a possibility to find the spatial distribution of particles, to calculate the cluster size and to obtain the temperature of the phase transition to the state under consideration. By means of this description it has been shown that cluster formation is possible in the system of attracting particles. The basic equation for the function of spatial distribution in the high-temperature limiting case, i.e., in the case of Boltzmann statistics, has been found. However, the dependence of the equilibrium size of the cluster on the thermodynamic limit has not been determined, and the dynamics of its formation has not been considered. This approach is correct also for various statistics when the problem to describe the systems like gas of interacting Fermi and Bose particles arises, in which spatial inhomogeneity of this type can appear.

Lately the interest on the Bose condensate of particles with negative scattering lengths has increased. The experiments [27–29] have shown that the condensate collapses when the number of particles reaches some critical value. The same result follows from the numerical solution of the Gross–Pitaevsky equation [30–32]. The collapse arises due to the tunneling through the barrier of particle attraction and quantum pressure that have given rise to the wave packet diffusion. Such situation can occur in the case of a shot-range attracting potential that can be described in terms of the scattering length. It is rather interesting to compare the stability of the model condensate with the long-range attracting potential, for example $1/R$ (gravitating gas), and the condensate with the short-range attracting potential (described by negative scattering length) by testing the solution of the Gross–Pitaevsky equation for stability. Perhaps, this model can be useful for investigating the early stages of the dynamically changing Universe [33]. Based on the statistical approach [3, 23], we describe the formation of a spatially inhomogeneous distribution in a system of interacting Bose particles. We obtain the conditions of cluster formation in the system for both Bose gas and condensate and describe the dynamics of cluster formation in the limiting case of high temperatures (Boltzmann statistics). We compare the properties of spatial inhomogeneity in the Bose condensate of particles with negative scattering lengths and particles with long-range attraction with respect to their instability to collapse.

The statistical description of Coulomb-like systems is one of the key problems of statistical physics too. Such systems might be helpful in testing the ideas concerning the description of systems with long-range interactions in terms of statistical mechanics. Solving the problem under consideration is complicated by the fact that standard methods of statistical physics cannot be used in the case of a system with Coulomb interaction. The structure formation in a system of colloidal particles or in dusty plasmas provides typical examples of a system with Coulomb-like interaction. One of the ways to describe the spatially inhomogeneous distribution in a system of interacting particles is to use the new method.

Many-particle systems with Coulomb interaction (Coulomb-like systems), such as plasmas, colloidal particles, electrolyte solutions, electron gas in solids, etc., are widely presented both in nature and under laboratory conditions. Many soft-matter systems, e.g., surfactant solutions, colloids in various solvents, and dust particles in plasmas, exhibit self-assembling into various structures. The interest in this system is generated by its applications to the studies of a variety of peculiar phenomena in various fields of science [34–36]. One of the most important problems here is the statistical description of Coulomb-like systems with high concentrations of interacting particles [37]. The formation of various crystal structures, transitions between crystalline phases of different symmetries, and melting-like phenomena [3, 38] are observed when concentration increases [39]. Moreover, dusty plasmas as well as colloidal suspensions may be perfect media for the experimental investigation of classical fluids and solids [39–48] since both direct measurements of the interaction between colloidal particles [49, 50] and theoretical treatment [51] reveal the Coulomb-like nature of the interaction in colloidal suspensions [35].

It is rather difficult to solve this problem by traditional methods of statistical mechanics since they cannot be applied to inhomogeneous systems with Coulomb-like interactions. In such cases, specific methods should be used taking into account the inhomogeneity of particle distributions. In particular, these methods should employ an appropriate procedure to find the dominant contribution to the partition function and to avoid free-energy divergences when the volume grows infinite. Only a few model systems of interacting particles are known for which the partition function can be evaluated exactly in the thermodynamic limiting case [52–54] and only a few results describing equilibrium states have been obtained within the framework of equilibrium statistical mechanics. In our approach, we propose that the known results can be obtained much more easily in terms of the method of collective variables and integral transformations [55]. Moreover, this method makes it possible to obtain the free energy of a classical plasma system in a regular manner up to an arbitrary order. A universal sequence of ordered structures was obtained from the description of the self-assembly using functional and statistical field theory [56].

The most interesting and exciting problem in condensed matter physics is the study of first-order phase transitions that produce states accompanied by the formation of a new phase with nonzero order parameter [57–65]. Such spatially inhomogeneous states can be thermodynamically stable with respect to small perturbations.

The relaxation towards a thermodynamically stable state is accompanied by the formation of critical nuclei of the new phase as the energy barrier that separates the nonequivalent minimum of the thermodynamic potential is overcome. All the characteristics of the first-order phase transition, i.e., the dimensions of the new phase bubbles, formation time, as well as time of relaxation towards a thermodynamically stable state, depend on the degree of this nonequivalence and the barrier [66, 69]. The required behavior of the free energy as a function of the order parameter can be realized either in the case of some special interparticle interaction in the system (in the microscopic treatment) [3, 7] or with the presence of an external field (in the phenomenological treatment) [67, 68, 70–76].

Interaction of the order parameter with the external fields of various nature [57–76] can modify the order of the transition and the intensity of such interaction determines the parameters of the new phase bubbles. In the general case, at both microscopic and phenomenological levels, the slowest subsystem is separated out while all features of the smaller-scale behavior are taken into account by averaging over probable fluctuations against the background of the order parameter selected. Choosing some certain order parameter facilitates the selection of probable states of the system at the microscopic level. From all probable microscopic states of the system, we select those that may be described in terms of the averaged order parameter whose variation scale is larger than the scale of probable fluctuations. The averaging results in the selection of probable states which may be described in terms of the selected order parameter associated with the "condensate" behavior of smaller-scale fluctuations. At the phenomenological macroscopic level, the first-order phase transition accompanied with the new phase bubbles formation also selects probable fluctuations of the order parameter with condensate behavior of the wavelength shorter than the bubble size. The formation time and relaxation towards a thermodynamically stable state depend first of all on the fluctuations of the scale smaller than that of the order parameter inhomogeneity. Thus, when describing the first-order phase transitions, it is of crucial importance to separate out the slowest subsystem and to select the states that contributes the most to the determination of the order parameter and the "condensate" correlated behavior of fluctuations. Only collective behavior of fluctuations of the scale smaller than the inhomogeneity of the characteristic variation of the order parameter can result in its destruction in the equilibrium case.

In order to study a specific physical system we have to first determine the dynamical quantities that should describe the system and in some way characterize both the systems and its interaction with the environment. A physical system is usually surrounded by some other system, e.g., the whole Universe. The environmental system is sometimes referred to as a reservoir, and in some cases it is called a thermostat. The thermostat is actually the key factor for the thermodynamic parameters of the system, e.g., temperature, volume etc.

Suppose the dynamical variables in an arbitrary point of space are determined by the variable that in the general case and in what follows is referred to as field. This variable is suitable to derive the expressions for thermodynamic quantities and to describe all probable changes occurring in the system. Such a field may imply, say, the occupation numbers, spin, dipole moment, as well as any other arbitrary quantity that describes specific properties of individual particles in an arbitrary point of the space. Moreover, the quantity under consideration is assumed to vary in space and to undergo all probable deviations from its average values (fluctuations). An important feature of the field is its geometric representation since it may be scalar, vector, or tensor of arbitrary rank. The geometric representation of the field is fairly important for the mathematical calculations of thermodynamic quantities and hence first of all we have to establish the physical characteristics of the system.

In what follows, we restrict the analysis to the scalar field functions and consider the peculiarities of the calculation procedure for other representations where needed. We consider specific systems. On the other hand, phase transitions in nonequilibrium systems, accompanied by the formation of dissipative structures, are strongly influenced by external noises associated mainly with the fluctuations of the governing parameter (external field). The structure of the nonequilibrium thermodynamic potential is entirely similar to that of the equilibrium potential whose governing parameter is temperature. For a structure, i.e., a spatially inhomogeneous distribution of the order parameter, to be formed in the nonequilibrium case, again the symmetry of the thermodynamic potential as a function of the governing parameter must be violated [24, 66, 77, 78]. It appears that the interaction of the system with the external noise — governing parameter fluctuations — would provide the desired asymmetric behavior of the thermodynamic potential so as to have nonequivalent minimums for different values of the order parameter.

In the case of noise-induced phase transitions, however, only the value of the governing parameter is changed, which is associated with the structure formation or phase transition. At the same time, the structure parameters depend, in the end, only on the degree of symmetry violation calculated in the traditional manner similar to the analysis of the thermodynamic potential nonlinearity [24–66, 68, 69]. In the case of nonequilibrium phase transitions, one can also trace the same approach as in the equilibrium description, with all probable fluctuations being averaged against the background of the separated order parameter. The separation of the order parameter is required for the experimental observation of the system and structural transformations occurring in it. The order parameter itself plays the role of a macroscopic parameter and the slowest variable in terms of which all the changes in the system can be described. However, a situation can occur, when the order parameter is not the slowest variable. An example of a nonequilibrium medium of this type is glass [79]. Equilibrium analogs of such systems exist as well. For example, the formation of a gas bubble in a superheated liquid, crystallization in a supercooled liquid, or superconductor in a magnetic field, for which the temperature or magnetic field can vary slower than the order parameter. The size of the second phase bubble depends on the observation temperature whereas the value averaged over various temperatures depends only on the character of interaction and on the dispersion of the governing parameter. Such cases are considered in our book.

The purpose of our study is to give a consistent description of phase transitions accompanied by the formation of new phase bubbles at both microscopic and phenomenological levels. Special attention is paid to the selection of states which contribute the most in the thermodynamic behavior of the system with inhomogeneous order parameter. Characteristic dimensions of the spatially inhomogeneous distribution of the order parameter may, in the end, be treated as a criterion for the selection of "condensed" fluctuations which make it possible to describe the system in terms of the order parameter.

The book introduces the readers to a consistent description of spatially inhomogeneous systems. First of all, we describe the statistical approach to the study of interacting systems with probable formation of spatially inhomogeneous particle distributions. Formation of such spatially inhomogeneous structures may be regarded as a first-order phase transition since it implies the formation of a new phase, i.e., a finite-size cluster whose properties differ from those of the initial homogeneous state. In terms of statistical approach, probable reasons for inducing spatially inhomogeneous particle distributions are indicated and the principle is formulated for the selection of states that provide the required behavior of the free energy in terms of the macroscopic order parameter.

Then the phenomenological approach is employed to show how the interaction of the separated order parameter with the fields of diverse nature can result in the change of the transition order; the parameters of the new phase formed are expressed in terms of the interaction intensity. Special attention is paid to the order parameter interaction with the field that varies slower than the order parameter. In this case, in a manner similar to the statistical analysis, the dimensions of the new phase of macrobubbles are expressed only in terms of the interaction parameters. Under the assumption that the field that is "external" with respect to the order parameter, fluctuates with proper dispersion, such noise-induced phase transition can be responsible for the formation of a new phase bubble with observed parameters.

Finally, a probability description is proposed for a noise-induced first-order phase transition under the assumption of fluctuation variations of the governing parameter. These can be the variations of the field that is "external" with respect to the order parameter and can simulate local variations of the microscopic interaction energy or temperature. Thus a possibility is provided to analyze, in terms of the general approach, the effect of fluctuations of any governing parameter on the stationary macroscopic behavior of the system. A special emphasis is made, in all sections of the book, on the description of the state-selection methods that yield a correct phenomenological expression for the thermodynamic potential in terms of the observed order parameter.

The self-gravitating systems and studying thereof have a fundamental physical background for testing the ideas concerning statistical and thermodynamic description of systems governed by long-range interaction. The statistical description of such systems is directly related to the problems of astrophysics [80, 81]. Some general problems at the field of self-gravitating systems have been studied for a long time [87, 88] and such systems seem to be very complicated when compared to other many-body systems. Normally, probable structure formation under various conditions can be described by the general methods.

The notion of equilibrium for the systems mentioned above is not always well defined and such systems exhibit nontrivial behavior with the occurrence of phase transitions associated with the gravitational collapse. The standard methods of equilibrium statistical mechanics cannot be used to study self-gravitating systems given that the thermodynamic ensembles are not equivalent: negative specific heat [89] in the microcanonical ensemble does not exist in the canonical description [90]. Two types of approaches (statistical and thermodynamic) have been developed to determine the equilibrium states of self-gravitating systems and to describe probable phase transitions [90, 91]. It is generally believed that the mean-field theory is exact for self-gravitating systems. According to the article [92] and the book [52] one can come to the conclusion that the thermodynamic limit of a self-gravitating system does not exist. Nevertheless, it is possible to take the usual thermodynamic limit and, consequently, to use safely the usual thermodynamic tools, primarily regularizing the long-range behavior of the gravitational potential and introducing a very large screening length. Only in this case, the system is thermodynamically unstable and the thermodynamic limit does exist [93].

Systems with long-range interactions, such as self-gravitating system, do not relax to the usual Gibbs thermodynamic equilibrium, but become trapped in quasi-stationary states whose lifetimes diverge as the number of particles grows. The theory that makes it possible to quantitatively predict the instability threshold of the spontaneous symmetry breaking for a class of d-dimensional self-gravitating systems was earlier presented in [94, 95]. Mostly due to the fact that self-gravitating systems exist in states far from equilibrium, the time of relaxation towards equilibrium state is very long. The homogeneous particle distribution in a self-gravitating system is not stable. The particle distribution in such a system initially is inhomogeneous in space. Therefore, the system brakes in a complex of inhomogeneous clusters that tend to collapse into a more condensed state. There were some attempts to include particle distribution inhomogeneity [3–5], but the solution has not yet been found. When the dependence of temperature on concentration is defined and, therefore, the concentration-dependence of pressure is defined too, one can obtain stable solutions for the gravitational formation of stars [90]. This approach seems to be somewhat inconsistent due to the fact that the equation of state should be obtained from the definition of the partition function though this definition for inhomogeneous systems is unknown [52]. Such inhomogeneous particle distribution, temperature, and chemical potential can be taken into account in the nonequilibrium statistical operator approach, where the possibility of a local change of thermodynamic parameters is considered [96]. The self-gravitating systems are nonequilibrium *a priori*.

In the present book we propose a new approach based on a nonequilibrium statistical operator [96] that is more suitable for the description of gravitation systems. The equation of state and all needed thermodynamic characteristics are defined by the equations that govern the largest contributions to the partition function. Thus, there is no need to introduce an additional hypothesis about the density-dependence of the temperature. This dependence is obtained by solving relevant thermodynamic relations that describe the extreme nonequilibrium partition function. The probable spatially inhomogeneous distributions of particles and temperature are obtained for simple cases.

In the equilibrium case, the well-known result [53, 54] for the partition function is reproduced. This approach is shown to be efficient to describe the inhomogeneous particle distributions and to find the thermodynamic parameters in a self-gravitating system. The main idea of this approach is to provide a detailed description of three-dimensional self-gravitating systems using the principles of nonequilibrium statistical mechanics and to obtain distributions of particles and temperature for fixed number of particles and energy within the system. We do not confine our inquiry to the dynamic aspects of this system, but describe possible inhomogeneous distributions of thermodynamic parameters for various external conditions.

So, in the present book we formulate an adequate mathematical method that would describe the formation and behavior of spatially inhomogeneous distributions of particles and fields in both equilibrium and nonequilibrium cases.

Chapter 1

Statistical Physics of Interacting Particle Systems

1.1 Systems of Particles with Interaction

The statistical and thermodynamic description of a system of interacting particles has been the basics for understanding condensed matter for many years. Significant success has been achieved as a result [80–82, 87, 88, 90–122]. Within the framework of this approach the phase transitions in a system of interacting particles from gas to liquid phases, and also from liquid to solid are described in [123–132]. Many models of the formation of separate phases in a system of interacting particles are considered depending on the value and character of the interparticle interaction [82, 97, 98]. Many systems may be treated as classical objects. These may be described by the classical Hamiltonian for N interacting particles, i.e.,

$$H = \sum_i^N \frac{p_i^2}{2m} + \sum_{i<j} V_{ij}, \tag{1.1}$$

where p_i is the momentum of the ith particle and V_{ij} is the potential energy of the interaction between the ith and j^{th} particles. The partition function of the system is given by

$$Z_N(V, P) = \frac{1}{N! h^{3N}} \int d^{3N} p \, d^{3N} r \exp\left(-\beta \sum_i^N \frac{p_i^2}{2m} - \beta \sum_{i<j} V_{ij}\right), \tag{1.2}$$

where $\beta = 1/kT$ is the reciprocal temperature. First of all, we integrate this expression over the momentum and thus obtain

$$Z_N(V,P) = \left(\frac{2\pi mkT}{h^2}\right)^{3N/2} Q_N, \qquad (1.3)$$

where

$$Q_N(V,P) = \frac{1}{N!} \int d^{3N}r \exp\left(-\beta \sum_{i<j} V_{ij}\right) \qquad (1.4)$$

is the configuration integral defined in terms of the energy. The general problem is to find this quantity. If the partition function is known, then we can find the expressions for all the thermodynamic functions of the system. The free energy of the system is given by

$$F = kTlnZ. \qquad (1.5)$$

Then for the entropy, we have

$$S = -\frac{\partial F}{\partial T}, \qquad (1.6)$$

and the internal energy is governed by the relation

$$U = F - T\frac{\partial F}{\partial T} = F + TS. \qquad (1.7)$$

The heat capacity of the system is described by the well-known formula

$$c = T\frac{\partial S}{\partial T} = \frac{\partial U}{\partial T}. \qquad (1.8)$$

With the configuration integral being known, we can calculate all thermodynamic characteristics of the system.

First of all, we shall show that in a quantum system without interaction there appear the "statistical interaction" that is responsible for the quantum nature of particles. Suppose, the natural interaction does not exist, $V_{ij} = 0$. We assume that this system of particles is not ideal gas, because allowance for the type of statistics leads to the appearance of the same interaction potential. The partition function of the system may be written as

$$Q_N(V, P) = Spe^{-\beta H} = \sum_n \left(\Phi_n e^{-\beta H} \Phi_n \right)$$

$$\simeq \frac{1}{N! h^{3N}} \int d^{3N} p d^{3N} q e^{-\beta H(p,q)}, \qquad (1.9)$$

where Φ_n is the wave function, and is determined by the full orthonormal system of eigenvalues of the Hamiltonian

$$H\Phi_n = \frac{1}{2m} \sum_i p_i^2 \Phi_n, \qquad (1.10)$$

that takes into account only the kinetic energy of each particle. When considering a quantum system we have to symmetrize or antisymmetrize the wave function depending on the kind of particles (Bose or Fermi). In the general case

$$\Phi_n = \frac{1}{\sqrt{N!}} \sum_i \delta_k \prod_k u_{k,p_i},$$

where $\delta_i = \pm 1$ for even or odd transformations corresponding to Fermi or Bose particles, respectively. The one-particle function may be written in the form

$$u_{k,p_i} = \frac{1}{\sqrt{V}} e^{\frac{i\mathbf{pr}}{\hbar}}.$$

In this case, the partition function may be obtained in the exact form given by

$$Spe^{-\beta H} = \frac{1}{N! h^{3N}} \int d^{3N} p d^{3N} q \exp - \left(\beta \sum_i \frac{p_i^2}{2m} \right) |\Phi_n|^2. \qquad (1.11)$$

Now we make use of the expansion

$$\sum_k \delta_k \prod_i [u_{k,p_i}] = 1 \mp \sum_{i,j} f_{ij} + \sum_{i,j,k} f_{ij} f_{ik} f_{kj} \mp \cdots,$$

where

$$f_{i,j} = f(\mathbf{r}_i - \mathbf{r}_j)$$

and

$$f(r) \equiv \exp\left(1 \mp \frac{\pi r^2}{\lambda^2}\right),$$

where

$$\lambda = \sqrt{\frac{2\pi\hbar^2}{mkT}}$$

is the thermal length of the de Broglie wave (the upper index corresponds to the Fermi statistics and the lower index corresponds to the Bose statistics) [53]. In the case $\mathbf{r}_i - \mathbf{r}_j \gg \lambda$, the partition function may be written in the form

$$Spe^{-\beta H} = \frac{1}{N!h^{3N}} \int d^{3N}p\, d^{3N}q \exp\left\{-\beta\left(\sum_i \frac{p_i^2}{2m} + \sum_{ij} V_{ij}\right)\right\}.$$
(1.12)

The statistically induced "interaction potential" can be written as

$$V_{ij} \equiv -kT \ln\left[1 \mp \exp\left(-\frac{2\pi|\mathbf{r}_i - \mathbf{r}_j|}{\lambda^2}\right)\right].$$
(1.13)

It is the first quantum correction to the partition function that takes into account the "effective potential" of the interaction between quantum particles. This potential possesses an attractive character for Bose particles and repulsive character for Fermi particles.

In the general case, the determination of the configuration integral is impossible except for the case of small ratio of potential to thermal energy that is suitable for the virial decomposition. In order to proceed to the general theory we should obtain the equation of states in the virial decomposition. For the case in point, we take the pair potential energy of interaction and determine the new function

$$\exp(-\beta V_{ij}) = 1 + f_{ij},$$
(1.14)

that has a presentation

$$\exp\left(-\beta \sum_{i<j} V_{ij}\right) = \prod_{i<j}(1 + f_{ij}).$$
(1.15)

If the distance between particles increases, or the temperature increases, the function f tends to zero very fast. If we have the small parameter given by the ratio of potential to kinetic energy (temperature) we can expand in the power series of the function f, i.e.,

$$\exp\left(-\beta \sum_{i<j} V_{ij}\right) = 1 + \sum_{i<j} f_{ij} + \sum_{i<j}\sum_{k<l} f_{ij}f_{kl}. \tag{1.16}$$

Given that function f depends only on the positions of particles i and j, we obtain

$$\int f_{ij}d^N r = V^{N-2}\int f_{ij}dr_i dr_j = V^{N-2}\int f(r)4\pi r^2 dr, \tag{1.17}$$

where r is the separation distance between particles. Having estimated several terms of the general presentation, we may present the configuration integral in the form

$$Q_N = \frac{1}{N!}V^N\left(1 + \frac{N}{V}b_2\right)^N, \tag{1.18}$$

where

$$b_2 = \frac{1}{2}\int_0^\infty f(r)4\pi r^2 dr$$

is the second virial coefficient. The equation of state that is associated with the configuration integral may be presented in the approximate form given by

$$P = \frac{NkT}{V}\left[1 - \frac{b_2}{v}\right], \qquad v = \frac{V}{N}, \tag{1.19}$$

In order to employ this equation to calculate the pressure, we have to know the explicit form of the interaction. In the case of spherical particles with weak attractive interaction, we have $V(r) = -V_0(r_0/r)^m$ if $r_0 \le r < \infty$ and $V(r) = 0$ if $0 \le r < r_0$, where r_0 is the particle size. We can calculate an approximate expression for

$$b_2 = \frac{a}{kT} - b,$$

where $b = (2\pi/3)r_0^3 = 4v_0$, and for

$$a = \frac{12}{m-3} V_0 v_0,$$

where v_0 is the particle volume. Then we note, that the equation of state reproduces the Van-der-Waals equation

$$\left(P + \frac{a}{v^2}\right)(v - b) = kT. \tag{1.20}$$

This formula solves the problem of how to describe the behavior of the system of particles with special interaction. In the case of a model system with Lennard–Jones interaction in the standard form

$$V(r) = V_0 \left\{ \left(\frac{r_0}{r}\right)^{12} - 2\left(\frac{r_0}{r}\right)^6 \right\}, \tag{1.21}$$

V_0 corresponds to the minimum value of energy for $r = r_0$. The second viral coefficient is determined by the integral

$$b_2 = \frac{2\pi}{3} \int_0^\infty \frac{1}{kT} \frac{dV}{dr} \exp\left(-\frac{V}{kT}\right) r^3 dr. \tag{1.22}$$

This integral may be obtained by expanding the expression in power series of the small parameter that represents the attractive part of the interaction. As a result, we have

$$b_2 = \frac{2\pi}{3} r_0^3 I\left(\frac{V_0}{kT}\right), \tag{1.23}$$

where

$$I(x) = \frac{1}{4} \sum_{n=0}^\infty \frac{2^n}{n!} \Gamma\left(\frac{2n-1}{4}\right) x^{\frac{2n+1}{4}}. \tag{1.24}$$

The temperature associated with the vanishing coefficient b_2 is referred to as Boile temperature. For this model, it is given by $T_B = 3,44(V_0/k)$. For low temperatures, the coefficient b_2 becomes negative.

For the case of hard molecules, the second virial coefficient has been calculated exactly [54]. It is given by

$$b_2 = v_0 \left(1 + \frac{\rho s}{v_0}\right),$$ (1.25)

where v_0 and s are the volume and the surface of the molecule, ρ is the average radius

$$\rho = \frac{1}{4\pi} \int \frac{R_1 + R_2}{2} d\omega.$$ (1.26)

Here R_1 and R_2 are principal radii of curvature, $d\omega$ is the surface element. This formula is correct for the case of molecules with arbitrary spatial orientations. The expression may be rewritten in the form

$$b_2 = 4v_0 f, \qquad f = \frac{1}{4}\left(1 + \frac{\rho s}{v_0}\right).$$ (1.27)

For the case of spherical molecules $f = 1$ and the second virial coefficient reduces to

$$b_2 = 4v_0.$$ (1.28)

For nonspherical molecules f is greater than one. For cylindrical or ellipsoidal molecules, the coefficient f increases if the length of the cylinder or the length of the major axis of the ellipsoid is increased. The rigorous analytical result for the coefficient f for an ellipsoidal molecule with rotational symmetry is given by

$$f = \frac{1}{4}\left\{1 + \frac{3}{4}\left[1 + \frac{\arcsin \varepsilon}{\varepsilon(1 - \varepsilon)^{1/2}}\left(1 + \frac{1 - \varepsilon^2}{2\varepsilon}\ln\frac{1 + \varepsilon}{1 - \varepsilon}\right)\right]\right\}, \quad (1.29)$$

where $\varepsilon = 0$ is the eccentricity of the ellipsoid. With the virial coefficient being known, we can obtain the equation of state and consider probable phases where this system of interacting particles can exist. First we have to show that the equation of state has a singularity. Considering the curve associated with this equation, we easily see that it has a minimum and a maximum. From the equation of state we have

$$P = \frac{kT}{v - b} - \frac{a}{v^2} \tag{1.30}$$

and thus we find the extreme parameters to be given by

$$\left(\frac{\partial P}{\partial v} \right)_T = -\frac{kT}{v - b}^2 - \frac{2a}{v^3} = 0. \tag{1.31}$$

In the maxima the second derivative should be equal to zero, i.e.,

$$\frac{\partial^2 P}{\partial v^2} = \frac{6}{v^4} - \frac{24ab}{v^5} = 0. \tag{1.32}$$

Having solved the two equations together we obtain the critical parameter associated with the instability. It is given by

$$P_c = \frac{a}{27b^2}, \qquad v_c = 3b, \qquad T_c = \frac{8a}{27kb}. \tag{1.33}$$

We introduce a dimensionless parameter

$$\tau = \frac{T}{T_c}, \qquad p = \frac{P}{P_c}, \qquad \nu = \frac{V}{V_c}, \tag{1.34}$$

then the dimensionless equation of state may be rewritten in the general form

$$\left(p + \frac{3}{\nu} \right) (3\nu - 1) = 8\tau. \tag{1.35}$$

This equation provides information about the specific features of the matter. Moreover, this equation of state may be used for all systems that satisfy the presented condition.

1.2   Models of Statistical Physics

In this section we present simple theoretical models that are used to describe systems with phase transitions. The most well-known model of this type is the Ising model.

Ising model. It represents a system of interacting particles allocated in the lattice with two probable orientations, up or down. The total number of probable configurations of this system is equal to 2^N, where N is the number of positions in the lattice. Each position may be attributed with the variable σ_x that takes two values ± 1 corresponding to two particle orientations, respectively. Each configuration is associated with the energy

$$H(\sigma) = \sum_{x,x'} I(x - x')\sigma_x\sigma_{x'} - h \sum_x \sigma_{x'}. \qquad (1.36)$$

It is usually accepted that the interaction involves only the nearest particles, i.e., particles that occupy the nearest positions, when $I(x) = 0$ for $x \neq a$, where a is the lattice constant. The Ising model is a simpler version of the model of magnetic substance, where $I(x)$ is the exchange interaction and h is the magnetic field. First of all, we consider the Ising model without magnetic field. The general state of a system with $I < 0$ satisfies the ferromagnetic ordering, all particle orientations are similar. The ground state is degenerate. The energy of this state takes the value $E_0 = N \sum_x I(x)$. For $I > 0$, the minimum energy occurs if the nearest particles have opposite orientations. As a result we have two lattices with different orientations. This structure represents the antiferromagnetic structure. Exact solutions in this model exist in the one- or two-dimensional cases. In the one-dimensional case the partition function may be written as

$$Z = \sum_{\sigma_i=\pm 1} \prod_i \exp(K\sigma_1\sigma_{i+1}), \qquad (1.37)$$

where the product is taken over all pairs of particles, and $K = I/2kT$. We introduce a matrix with elements $P(\sigma, \sigma') = \exp(K\sigma\sigma')$. Given $\sigma = \pm 1$, we have

$$P(\sigma, \sigma') = \exp(K\sigma\sigma') = \cosh K + \sigma\sigma' \sinh K. \qquad (1.38)$$

This relation is exact. We select σ that satisfies this. For $\sigma_i^2 = 1$ and $\sum_{\sigma_i \pm 1} = 0$ the odd terms of the series over the parameter σ tend to zero, and we only have even terms. Thus the partition function is given by

$$Z = (2\cosh K)^N + (2\sinh K)^N. \qquad (1.39)$$

The same result may be obtained in terms of the matrix method by considering the eigenvalues

$$P(\sigma, \sigma')\alpha(\sigma) = \lambda\alpha(\sigma) \qquad (1.40)$$

or the obvious expression

$$\begin{bmatrix} e^K & e^{-K} \\ e^{-K} & e^K \end{bmatrix} \begin{bmatrix} a\,(1) \\ a\,(-1) \end{bmatrix} = \lambda \begin{bmatrix} a\,(1) \\ a\,(-1) \end{bmatrix}. \qquad (1.41)$$

This secular equation yields

$$\lambda = e^K \pm e^{-K} = \begin{cases} 2\cosh K \\ 2\sinh K \end{cases}, \qquad (1.42)$$

and thus we obtain the previous result for the partition function. This method is also correct in the case, with the magnetic field. In this case, we have to introduce an additional matrix

$$Q = \begin{bmatrix} e^C & 0 \\ 0 & e^{-C} \end{bmatrix}, \qquad (1.43)$$

where $C = h/kT$, and to consider the problem of eigenvalues of the matrix QP. In this case, for the maximum eigenvalue we obtain

$$\lambda_{\max} = e^K \cosh C + (e^{2K} \sinh^2 C + e^{-2K}) \qquad (1.44)$$

and can calculate the partition function from the formula $Z = \lambda_{\max}^N$. After that, we find the magnetization coefficient to be given by

$$\mu = \frac{\sinh C}{\sinh^2 C + e^{-4K}}. \qquad (1.45)$$

The result thus obtained clearly shows that the partition function of the one-dimensional Ising model has no singularities and is not associated with any phase transitions in this interacting system.

In the two-dimensional case, we have to calculate the partition function

$$Z = \sum_{\sigma_i = \pm 1} \prod_i \exp(K\sigma_i\sigma_j). \qquad (1.46)$$

Moreover, we begin to consider particles that occupy positions in the two-dimensional lattice. Then, we introduce the total number of pairs

S and the coordination number z bearing in mind that $S = zN/2$ and introduce the quantity $x = \tanh K$. Thus, we obtain the partition function as a power series of x given by

$$Z = (\cosh K)^S \sum_{\sigma_i = \pm 1} \prod_i (1 + x\sigma_i\sigma_j). \qquad (1.47)$$

In the next step, we introduce the quantity Ω_n, i.e., the number of closed polygons with the unit length n. Then the partition function reduces to

$$Z = 2^N (\cosh K)^S \left(1 + \sum_n \Omega_n x^n \right). \qquad (1.48)$$

This partition function may be obtained in a different way, by introducing the number l of pairs with opposite directions of particle orientations. Let S be the total complete number of such pairs, then we have

$$\sum_{i,j} \sigma_i\sigma_j = (S - l)(1) + l(-1) = S - 2l. \qquad (1.49)$$

Using these terms, we can present the statistical sums of the form

$$Z = 2^N e^{SK} \left(1 + \sum_l \omega_l e^{-2rK} \right), \qquad (1.50)$$

where ω_l is the number of configurations containing l pairs of particles with opposite directions of orientation. We have two equivalent presentations of the partition function for the two-dimensional Ising model. We do not consider the rigorous exact solution of this model that is discussed in detail in many books. We just note that in this two-dimensional case there exists a phase transition of the second order and all parameters of this phase transition are well-known. A useful generalization is the continual prototype of the Ising model. In this version, a value of the continuum variable $\varphi(x)$ is set in the center of the lattice, that can be changed from $+\infty$ to $-\infty$. The Hamiltonian of this system takes the form:

$$H = \frac{I}{2} \sum_{x,b} \left[\varphi(x) - \varphi(x + b) \right]^2 + \lambda \sum_x \left[\varphi^2(x) - \varphi_0^2 \right], \qquad (1.51)$$

where b is the lattice constant. This model possesses isomorphism with the Ising model insofar as in the limiting case $\varphi(x)$ takes two values $\pm\varphi_0$. For positive λ and I, the ground state of the system is doubly degenerate. In a weakly excited state $\varphi(x)$ slowly changes. The continuum analog of this model is given by

$$H = \frac{I}{2} \int \left\{ \frac{c}{2}(\nabla\varphi)^2 + \lambda'(\varphi^2 - \varphi_0^2)^2 \right\} dx, \tag{1.52}$$

where $c = Ib^{2-d}$, $a = -2\lambda\varphi_0^2 b^{-d}$, $\lambda' = \lambda b^{-d}$, d is the dimension of the space. It may be shown that in the case $d > 2$, a phase transition exists in the system. However, no exact solution of the continuum model has been found up to now.

Model of hard solution of substitution

In order to describe the hard solution of substitution we use the model starting with the lattice occupied by different particle species. The interaction between particles is only of pairwise character, though different for different particle species. In this model the Hamiltonian of the system may be written in the form:

$$H = \frac{1}{2} \sum_{r,r'} \left\{ \tilde{V}_{aa} n_a(\mathbf{r}) n_a(\mathbf{r}') + \tilde{V}_{bb} n_b(\mathbf{r}) n_b(\mathbf{r}') + 2\tilde{V}_{ab} n_a(\mathbf{r}) n_b(\mathbf{r}') \right\}, \tag{1.53}$$

where n_i is the occupation number for the particle species a or b, and $V_{ij}(r, r')$ is the energy of interaction between particles of species a and b. Each occupation number n_a takes the value 1 if the position is occupied by a particle of species a and 0, if this position is occupied by a particle of other species. The function n_i strictly satisfies the relation $n_a(\mathbf{r}) + n_b(\mathbf{r}) = 1$. This relation makes it possible to rewrite the Hamiltonian as given by

$$H = H_0 + \frac{1}{2} \sum_{r,r'} \tilde{V}(\mathbf{r}, \mathbf{r}') n(\mathbf{r}) n(\mathbf{r}'), \tag{1.54}$$

where the notation $n_a(\mathbf{r}) \equiv n(\mathbf{r})$ is introduced and n_b is excluded in terms of the previous relation. Moreover,

$$\tilde{V}(\mathbf{r}, \mathbf{r}') = \tilde{V}_{aa}(\mathbf{r}, \mathbf{r}') + \tilde{V}_{bb}(\mathbf{r}, \mathbf{r}') - 2\tilde{V}_{ab}(\mathbf{r}, \mathbf{r}')$$

and

$$H_0 = \frac{1}{2} \sum_{r,r'} \left\{ \tilde{V}_{bb}(\mathbf{r}, \mathbf{r}') \left[1 - n_a(\mathbf{r}) - n_a(\mathbf{r}') \right] + 2\tilde{V}_{ab}(\mathbf{r}, \mathbf{r}') n_a(\mathbf{r}) \right\}.$$

(1.55)

Within the context of the relations

$$\sum_{r'} \tilde{V}_{bb}(\mathbf{r}, \mathbf{r}') = \tilde{V}_{bb}(r) = V_{bb}(0),$$

$$\sum_{r'} \tilde{V}_{ab}(\mathbf{r}, \mathbf{r}') = \tilde{V}_{ab}(r) = V_{ab}(0),$$

$$\sum_{r} n_a(r) = N_a$$

we obtain

$$H_0 = \frac{1}{2} \tilde{V}_{bb}(0) \left[N - 2N_a \right] + N_a \tilde{V}_{ab}(0).$$

(1.56)

Given that H_0 is a finite quantity and does not depend on the positions of particles, we can use this value as the reference energy and the Hamiltonian of the system may be presented in the standard form given by

$$H = \frac{1}{2} \sum_{r,r'} \tilde{V}(\mathbf{r}, \mathbf{r}') n(\mathbf{r}) n(\mathbf{r}').$$

(1.57)

The free energy of the system may be found in this case by the method of statistical thermodynamics, we write it in the form

$$F = \frac{1}{2} \sum_{r,r'} \tilde{V}(\mathbf{r}, \mathbf{r}') n(\mathbf{r}) n(\mathbf{r}') + kT \sum_{r} [n(\mathbf{r}) \ln n(\mathbf{r}) - (1 - n(\mathbf{r}))$$

$$\ln(1 - n(\mathbf{r}))] - \mu \sum_{r} n(\mathbf{r}).$$

(1.58)

In the disordered state $n(r) = c$, where c is the relative concentration, the occurrence of concentration heterogeneity may be considered in terms of an additional part $n(r) = c + \phi(r)$, then the free energy may be expanded in power series of this parameter. The first term contained in the expression for the free energy may be written as

$$F = \frac{1}{2} \sum_{r,r'} \left\{ \tilde{V}(\mathbf{r}, \mathbf{r}') + \frac{kT}{c(1-c)} \delta_{\mathbf{r}, \mathbf{r}'} \right\} \phi(r)\phi(r'). \tag{1.59}$$

The discontinuity of the concentration $\phi(r)$ may be presented as a superposition of flat waves and expanded in Fourier series, i.e.,

$$\phi(\mathbf{r}) = -\frac{1}{2N} \sum_q \phi(\mathbf{q}) \exp(i\mathbf{rq}).$$

Substituting this presentation in the previous expression yields the free energy given by

$$F = -\frac{1}{2N} \sum_q b(\mathbf{q})|\phi(\mathbf{q})|^2, \tag{1.60}$$

where

$$b(\mathbf{q}) = V(\mathbf{q}) + \frac{kT}{c(1-c)}. \tag{1.61}$$

If the minimum value $b(\mathbf{q})$ vanishes or is negative, then the system remains in the unstable state accompanied with the formation of a static concentration wave with $q = q_0 \neq 0$. The condition $\min b(q, T_c) = b(0, T) = 0$ determines the temperature of the spinodal decomposition in the system,

$$T_c = \frac{c(1-c)V(0)}{k}.$$

This result can be obtained in other ways. From the free energy, we obtain an equation for the self-consistent mean field that yields the general expression for the distribution function. Thus, we have

$$n(\mathbf{r}) = \left\{ \exp\left[-\frac{\mu}{kT} + \sum_{r'} \frac{V(\mathbf{r}, \mathbf{r}')}{kT} n(\mathbf{r}') \right] + 1 \right\}^{-1}. \tag{1.62}$$

The latter equation is a nonlinear integral equation and has many solutions. Each solution corresponds to a minimum of the free energy. Each set of thermodynamic parameters is associated with the solution of this equation corresponding to an absolute minimum of the free energy. The transition from one solution to another describes a phase transition. The temperature of the phase transition may be determined as the branching point in the equation for the self-consistent mean field. The equation of branching may be obtained by expanding the equation for the self-consistent mean field over the variations of the distribution function. We substitute $n(r)$ on $n(r) = c + \delta n(r)$ in the first-order term of the expansion and thus obtain a new equation that is given by

$$\delta n(\mathbf{r}) = -\frac{c(1-c)}{kT} \sum_{r'} V(\mathbf{r}, \mathbf{r}') \delta n(\mathbf{r}'). \tag{1.63}$$

We multiply the left- and right-hand parts of this equation by $\exp(-iqr)$ and sum up by r. Thus we have

$$\delta n(\mathbf{q}) = -\frac{c(1-c)}{kT} V(\mathbf{q}) \delta n(\mathbf{q}). \tag{1.64}$$

This equation has a nontrivial solution $\delta n(\mathbf{q}) \neq 0$ if

$$1 = -\frac{c(1-c)}{kT} V(\mathbf{q}) \quad \text{or} \quad T = -\frac{c(1-c)}{k} V(\mathbf{q}).$$

The region of changing temperatures is determined by the inequality $0 < T(q) \leq T_c$, where

$$T_c = -\frac{c(1-c)}{kT} \min V(\mathbf{q}).$$

The model of hard solution of substitution provides a possibility to describe the condition for the decomposition of the homogeneous distribution of particles of different species and to find the temperature of the phase transition with the formation of a periodic distribution of these particles.

Heisenberg model

The description of a system of interacting particles in terms of the Ising model is justified in the case of very strong anisotropy of the properties of the condensed matter. Another extreme case represents the isotropic Heisenberg model. This model describes the exchange interaction between electrons localized in the center of the lattice.

We introduce the spin operator $S(\mathbf{x}')$ at each point $\mathbf{x}$ of the lattice. The commutator for the spin components takes the form

$$[S_i(\mathbf{x}), S_k(\mathbf{x}')] = i\varepsilon_{ikl}S_l(\mathbf{x})\delta_{x,x'}. \qquad (1.65)$$

The Hamiltonian of the system of spins is given by the standard expression

$$H = \sum_{\mathbf{x},\mathbf{x}'} I(\mathbf{x} - \mathbf{x}')S(\mathbf{x})S(\mathbf{x}') - \mathbf{h}\sum_{\mathbf{x}} S(\mathbf{x}). \qquad (1.66)$$

In contrast to the Ising model, individual components $S_i(\mathbf{x})$ are not conserved in this model, while the square of the moment of the system M^2 and one component M_z are conserved, i.e., $\mathbf{M} = \sum_{\mathbf{x}} S(\mathbf{x})$.

Let us consider the case of a ferromagnet, $I < 0$, without magnetic field $h = 0$. The ground state of this system corresponds to the maximum probable value of M^2. All spins are oriented in the same direction, say, along the z-axis, and the projection of the spin S_z has the maximum probable value. The projection of the total moment is given by $M_z = NS_z$, where N is the number of particles. The smallest of the excited states is the spin wave. The energy of the spin wave depends on the square of the wave vector, $\mathbf{k}^2$, for small $k < 1/a$, where a is the lattice constant. If the number of spin waves is small, then the number of quasi-particles may be described in terms of the Bose statistics (their appearance changes the total spin). For low temperatures ($T < T_c$), the number of excited states within a unit volume is equal by the order of magnitude to the product of the phase volumes $(4\pi/3)p_T^2$ and cell volumes $(2\pi)^3$. The thermal momentum p_T may be estimated from the relation $Ip_T^2 \sim T$. As a result, we obtain a relation for the temperature-dependence $M_z(T)$, i.e.,

$$M_z(T) = M_z^0\left[1 - \left(\frac{T}{T_c}\right)^{3/2}\right], \qquad T < T_c. \qquad (1.67)$$

For $T < T_c \sim I$ the long-range order in the alignment of spins is conserved. For $T \gg T_c$, the spin interaction is unimportant and the long-range order does not occur. Thus, we have a phase transition for $T \sim I$ and $h = 0$. This result is correct in the three-dimensional case. In the cases of two or one dimensions the number of spin waves that can be calculated in terms of the Plank formula is divergent. Calculations in two and one dimensions show that phase transitions in all these cases do not occur. The ground state of an antiferromagnet in the Heisenberg model $I > 0$ has a singularity. Exactly indisputable proof of the existence of two lattices with different spin orientations in the three-dimensional case has not yet been obtained. This system possesses the property that below and above the point of the phase transition in a magnetic field there appears a nonzero magnetic moment along the direction of the field h. The magnetic susceptibility is finite for any temperature. The order in this system is not induced by the magnetic field because the phase transition does not disappear for finite magnetic fields h, but the temperature is changed. For very strong anisotropic properties of the type of easy axis all moments have similar directions. In this case we can use the Hamiltonian

$$H = \sum_{\mathbf{x},\mathbf{x}'} I(\mathbf{x} - \mathbf{x}')S_z(\mathbf{x})S_z(\mathbf{x}') - h_z S_z, \qquad (1.68)$$

that in the case of spin $1/2$ is in agreement with the result of the Ising model. For strong anisotropic properties of the easy-plane type the Hamiltonian of a flat magnetic may be written in the form

$$H = \sum_{\mathbf{x},\mathbf{x}'} I(\mathbf{x}' - \mathbf{x}')\left\{ S_x(\mathbf{x})S_x(\mathbf{x}') + S_y(\mathbf{x})S_y(\mathbf{x}') \right\}. \qquad (1.69)$$

This Heisenberg Hamiltonian corresponds to the model of a flat magnetic object where the point of phase transition exists in the three-dimensional case.

1.3   The Model of Hard Spheres with Attractive Interaction

The classical density-functional theory has become increasingly popular in the study of liquid–solid transitions [82, 97]. The popularity of these methods is at least in part due to their success in providing an accurate description of the transition from a fluid to a close-packed-solid in a system of hard spheres [82, 98]. Within the context of these results, it seems that the density-functional theory works well, at least for systems of particles with "hard" interactions for which the equilibrium solid phase is closely packed [82, 98, 99]. The success of the density-functional theories in the case of hard-sphere potentials extends as well to systems of particles with softer interactions, where the thermodynamically stable solid phase may not be closely packed. The density-functional theory applied to nonuniform classical liquids has been able to reveal a wide range of physical properties of simple solid systems [123–132]. It is well known that hard spheres are used as the usual reference system. In order to make the system more realistic, however, we have to include some attractive interactions. Accordingly, considerable effort has gone into developing a free energy function that would describe nonuniform hard-sphere systems and at present there exists a quite good functional approach for these cases. We note that there are no current equations of state that would allow for hard-disk and hard-sphere fluids in the close-packing limiting case [105]. A new problem arises when the density function is used to describe less packed crystal structures, such as the *bbc* lattice whose state is never an equilibrium, state for hard spheres though it still may be useful when hard spheres provide a reference system for softer interactions. The basic solid-state property has not yet been fully formulated, nevertheless the density-functional theory is very well adapted to the high-density equation of state [99, 104, 105]. The essence of the approach is to incorporate physical information about the static solid into the original density-functional theory model based on the fluid state.

The problem regarding formulating solid-state conditions for given time has not yet been fully solved. It is associated first of all with the thermodynamic limit in the statistical approach of the description of interacting particles that is not quite applicable. Here we have to emphasize two aspects. The first one is the statistical description with a thermodynamic limit in a system that cannot pass through non-allowed states with fixed numbers of particles and concentrations. With the number of particles being fixed, the particle density is shown to increase in the course of condensation of the substance, thus resulting in a decrease of the volume of the system. Another problem is that the system of interacting particles tends to the crystal condition after passing through many intermediate conditions, such as a colliding liquid. The description of a colliding-liquid state causes much difficulties both in computing and of basic characterics. The most important for this research is that there remains the problem of reduction of probable states of the systems in its statistical description. We can always describe the criteria of instability in a system, but we cannot specify the final state. A considerable fraction of the elements, when compressed, undergo the first-order phase transition from a disordered state (gas, fluid, or liquid) to a face-centered cubic (fcc) or hexagonal closely packed (hcp) crystal. Both computer simulations and experiments show that this freezing transition is already apparent in the simple model of interacting particles [98, 100–102]. The theory can explain only the transition "afterwards". A thorough theory of hard sphere freezing should therefore specify this property of the partition function that leads to the observed symmetry breaking. Therefore, the approach seems to be more reasonable when the final condition is given *a priori*. Then, proceeding with this condition and basing on some principles of the statistical description, we obtain a criterion for such a condition to be realized. In this plan, the description of a simple model of a system of interacting spheres is offered with final energy of repulsion during contact and far attraction of gravitation. The particle systems with purely attractive gravitation interaction are known to exhibit collapse, sometimes called a zero-order phase transition [103, 127]. If no short-range cutoff is introduced, then the discontinuous jump is infinite. This makes all normal uncollapsed states of the self-attractive system metastable with respect to such a collapse. On the other hand, if a similar form of the short-range cutoffs is introduced, then both entropy and free-energy jumps are finite. In this case, as a result of the collapse, the system goes into a nonsingular state with a dense core. The real nature of these states depends on the details of the short-range behavior of the potential. Such a model system of interacting particles is of interest in the

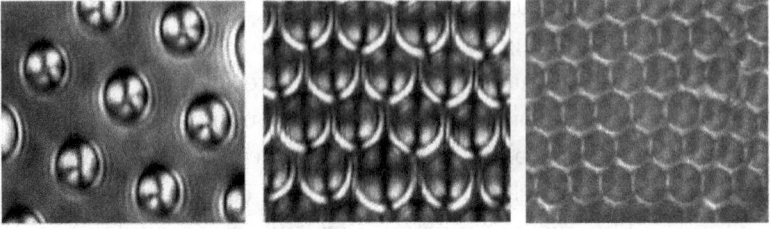

Fig. 1.1.   Possible transmutation of structures by Coulomb-like attractive inter-action in the system of glycerol droplets [184].

study of final states for the gravitation of particles with hard cores, as well as for the applications to colloidal particles in liquid crystals Fig. 1.1.

The second aspect is that for systems with long-range attraction and short-range repulsion, the stages of formation of a crystal state may be accompanied by changes of packing. In most crystal structures, changes occur that are associated with changes of packing (for example, cubic can be transformed into hexagonal). If we imagine that it can also change its form of the structural unit (for example, the volume of a deformed spherical colloidal particle) under strong attraction, then it is quite probable that there occurs a phase transition with the change of packing. For such systems, we propose the description given below. It is however very important that even for classical particles we have to apply the Fermi statistics because two classical finite-size particles cannot occupy the same point in space. This fact on the description of phase transitions in the crystal state has not yet obtained the due consideration [133].

Let us consider the model of a system of interacting particles in a lattice with the potential

$$W(\mathbf{r} - \mathbf{r}') = -\frac{Q^2}{\mathbf{r} - \mathbf{r}'} + U\theta(2R_0 - \mathbf{r} - \mathbf{r}'). \qquad (1.70)$$

The first term presents the attraction-type gravitation energy, the second term presents the repulsive energy of interaction between two particles separated by a distance of $2R_0$, where R_0 is the radius of a particle, $\theta(2R_0 - \mathbf{r} + \mathbf{r}')$ is the step function. The free energy of such a system may be written in the form:

$$F = \frac{1}{2} \sum W(\mathbf{r}, \mathbf{r}') f(\mathbf{r}) f(\mathbf{r}')$$
$$+ kT \sum \left\{ f(\mathbf{r}) \ln f(\mathbf{r}) - [1 - f(\mathbf{r})] \ln[1 - f(\mathbf{r})] \right\}. \quad (1.71)$$

Here $U(\mathbf{r} - \mathbf{r}')$ is the interaction potential, $f(\mathbf{r})$ is the probability that the particles fill the initial arbitrary cellular lattice. The whole space is divided into cells, only one particle can occupy each cell. It should be noted that the entropy part of the free energy is of Fermi type and that two particles cannot occupy the same point in space. The thermodynamic functions of the state correspond to a solution that describes some phase of particle arrangement. If their distribution can be inhomogeneous, then the solution serves to find the stable phase associated with the character, interaction, and temperature. If the particle solution is disordered, then by definition the mean value would be $f(\mathbf{r}) = f_0$, where f_0 is the relative particle concentration. The concentration inhomogeneity gives rise to an additional term $f(\mathbf{r}) = f_0 + \Delta f(\mathbf{r})$, where $\Delta f(\mathbf{r})$ is the change of the probability distribution function of particles. In the density-functional theories, the solid density is given *a priori* by the Gaussian parametrization

$$f(\mathbf{r}) = \left(\frac{\pi}{a} \right)^{3/2} \sum_R \exp\left(-a|\mathbf{r} - \mathbf{R}_i|^2 \right),$$

where the sum is taken over the real lattice vector $\mathbf{R}_i$ that depends on the particular crystal structure, and parameter a is the order parameter that can be different for liquid and solid states, δ-function density profile in the Fourier space and may be expressed as

$$f(\mathbf{r} = f_0 + \sum \Delta f(\mathbf{q}) \exp(i\mathbf{q}\mathbf{r}),$$

where the sum is now over the reciprocal lattice vector $\mathbf{q}$. In our case

$$f_0 = \frac{N}{N_0} = \alpha v = \frac{8}{\pi} v,$$

where v is the packing factor. It should be noted that the packing factor and its coefficient represent a function of particle distribution in cells depending on the cell shape. This representation corresponds to square cells with linear sides equal to the particle diameters. In other cases, the coefficient before the packing factor can take other values. For large values of a, it is typically found in solids. In particular, in the case of a homogeneous particle distribution in the specified space, we obtain a formula for an ideal free energy given by

$$F(f_0) = \frac{1}{2}N^2\left(Uv^2 - \frac{3W_G v^{1/3}}{2N^{1/3}}\right)$$

$$+ kT\left\{\ln(\alpha v) + \frac{1 - \alpha v}{\alpha v}\ln(1 - \alpha v)\right\}. \qquad (1.72)$$

The equation of state may be represented as

$$P = -\frac{\partial F}{\partial V}\bigg|_{N,S} = -\frac{\partial F}{\partial v}\frac{\partial v}{\partial V} \qquad (1.73)$$

and in our case $-\partial v/\partial V = v/V$, we obtain the equation of state in the form

$$PV = \frac{\partial F}{\partial v}v = \frac{1}{2}N^2\left(Uv - \frac{W_G v^{1/3}}{2N^{1/3}}\right) - \frac{kTN}{\alpha v}\ln(1 - \alpha v). \qquad (1.74)$$

When $f_0 = \alpha v \leq 1$, we obtain the equation of state for the cellular gas with the term associated with the interaction in the system, i.e.,

$$PV = kTN + \frac{1}{2}N^2\left(Uv - \frac{W_G v^{1/3}}{2N^{1/3}}\right). \qquad (1.75)$$

This equation reproduces the Van-der-Waals type equation of state for hard spheres, where the second term represents the interaction in this system. This result provides a very good approximation for an equation of state for systems with all models of the interaction potential with a repulsive hard core and finite-range attraction. Thermodynamic properties of elastic hard-sphere systems depend on the temperature in a trivial manner. Two methods can be used to calculate the pressure in the packing. The first method consists of calculating the rate h that then yields the pressure by the formula

$$\frac{PV}{kTN} = 1 + \frac{h}{h_0},$$

where

$$h_0 = 8\sqrt{\frac{\pi \langle v^2 \rangle}{3}}$$

and $[R^2 N(N-1)]/V$ is the low-density collision rate for greater packing of hard spheres, $\langle v^2 \rangle$ is the mean square velocity proportional to T. The second approach is based on the fact that the equation of state of hard spheres is related to the radial distribution function $g(r)$ where $r = 2R$, $(PV)/(kTN) = 1 + 4vg(2R)$. The two methods give close values, but the second one requires some caution, the correlation distribution function is difficult to measure precisely because the radial distribution function can rise or fall rapidly close to $r = 2R$ [106]. This equation of state is not equivalent to the other equation of state where the particle volume is excluded

$$PV = kTN \frac{1 + v + v^2 - v^3}{(1 - v)^3}$$

while the asymptotic behavior of the system is given correctly. When

$$\frac{W_G}{2U} \geq N^{1/3} v^{2/3} = N \left(\frac{R}{R_0} \right)^2,$$

where R is the size of the system, and in the case when

$$\frac{R}{R_0} \leq N \left(\frac{W_G}{4kT} \right),$$

a collapse can occur in the system that is probable for $v = (4kT/W_G)^3 N^{-2}$. In the general case, there is a solution for the equation of state, when the pressure in the system becomes negative, thus attraction dominates and it forms the closely packed structure, i.e.,

$$\frac{1}{2} N^2 \left(Uv - \frac{W_G v^{1/3}}{2N^{1/3}} \right) - \frac{kTN}{\alpha v} \ln(1 - \alpha v) = 0. \tag{1.76}$$

In the case when $f_0 = \alpha v \to 1$ we have

$$\alpha v = 1 - \exp\left\{ \frac{1}{2\alpha}\left[\frac{NU}{kT} - \frac{W_G \alpha^{1/3} N^{2/3}}{2kT} \right] \right\}, \qquad (1.77)$$

that is probable when the average distance between particles $l \leq (W_G/2U)R_0$.

When passing to the continuum description, we can write the free energy increment associated with the inhomogeneous particle distribution in terms of the power series expansion by using the long-wavelength expansion of the concentration. In this case, we may rewrite the change in the free energy in the form

$$F = \frac{N}{2V}\sum\left\{ -\frac{4\pi Q^2}{q^2}\left(\delta_{ij} - \frac{q_i q_j}{q^2}\right) + U\frac{4\pi}{3}R_0^3 + \frac{kT}{\alpha v(1 - \alpha v)} \right\}$$
$$\times \Delta f(\mathbf{q})\Delta f(-\mathbf{q}), \qquad (1.78)$$

where $\Delta f(\mathbf{q})$ is the Fourier transformer function of the distribution of particles in the cells. For the hexagonal packing, we have the solution in the form

$$qR_0 = \left(\frac{3W_G}{4U}\right)^{1/2}\left\{ 1 + \frac{kT}{U\alpha v^2(1 - \alpha v)} \right\}^{-1/2} \qquad (1.79)$$

and, when the $kT \to 0$, we have

$$qR_0 = \left(\frac{3W_G}{4U}\right)^{1/2}. \qquad (1.80)$$

This result is opposite to the structure with the wavelength shorter than twice the size of a cell. It is possible for $(cW_G)/U \geq 4\pi^2/3$. Thus the attraction obviously dominates over repulsion and can result in the deformation of a particle with a soft core. The shape of a cell can vary from the initial cubic cell to the hexagon. The conditions of such transformation can be found only due to physical reasons. Attraction between particles provokes a collapse in the system. The particles begin to press upon each other and finally violate the equilibrium condition that leads to the deformation of the particles. A final state in such a system occurs when all particles occupy a volume equal to the sum of volumes of all the particles. But this does not change the volume of an individual particle but only changes the shapes of these particles. For liquid particles, it is possible to estimate the pressure produced by the gravitational attraction and to compare it to the Laplacian pressure in a separate liquid particle. We shall carry out the estimation using $W_G/\sigma R^2 \geq 4/m$, where σ is the surface energy and m is the number of probable nearest neighbors. We can count the number of nearest neighbors within the context of the following reasoning. We draw a sphere of size $d = 2R$. On the surfaces of this sphere, we can place m circles with the area πR^2. Thus, we find that for a spherical particle the number of probable nearest neighbors $m = 16$. If the particles are deformed, then the distance between their centers should be less than d, thus we obtain the condition $4\pi^2/3 \geq 4/m$, that quite satisfies the condition of deformation for such a structural unit with a soft core.

In the case

$$\frac{kT}{U\alpha v^2(1 - \alpha v)} \geq 1,$$

we obtain

$$\lambda = \pi d \left[\frac{4kT}{3W_G\alpha v^2(1 - \alpha v)}\right]^{1/2},$$

the period of the particle distribution in the case of high temperatures. It is quite probable to consider a structure when the attractive interaction is very significant and to realize the condition of only equilibrium distribution of particles in the case when the repulsive interaction is negligible and the kinetic energy provides a stabilizing factor in the positioning of particles. In the case of an emulsion particle with the interactions of Coulomb screening, it is probable to observe a change of the structure with the changes of the shapes of droplets immersed in the fluid medium. This phenomenon can be observed in a system of colloidal particles with soft cores.

Thus, the behavior of interacting particles with long-range attraction and short-range repulsive interaction can be described for all stages of existence [3]. Beginning from the gas state the attraction results in the condensation of the system. In the condensed matter under the temperature decrease the attraction can induce a transition to the solid state with the packing factor of cubic structure. For some relation between the attraction in the system and the repulsion interaction that is produced by the finite volume of the structural unit there can occur a new structural transition that changes the structural factor. The cubic lattice is transformed to the hexagonal structure. Thus the shape of the structural unit can be changed. For example, a sphere can be transformed to a hexagon (or a disk in the hexagonal cell) without change of volume. Thus the initial volume is equal to the sum of individual volumes of separate structural units. Thus, even the collapse in the system, that provokes strong attraction between particles, should lead to the formation of a new structure with the increase of the structural factor up to one, and this state is final in the system with strong long-range attraction and short-range repulsion of the soft type.

1.4   Nonideal Gas at Low Temperatures

The nonideal gas is a rarefied system of interacting particles with a finite-range potential. In this case, the interaction may be treated as a perturbation of an ideal gas. The consideration of the nonideal gas is the next step to the understanding of the models of realistic physical systems. For low temperatures, two parameters with length dimensions are very important — the thermal length λ and the mean distance between particles $v^{1/3}$. These lengths can be equal but should be longer than the range of action of the interaction potential. It is well-known from quantum mechanics that the dispersion of particles does not depend on the form of the potential but is determined by a single parameter a that is called the dispersion length. If the interaction potential is not equal to the potential of hard spheres, has finite range of action, and does not form intermediate states, then the Hamiltonian of the system of N

interacting identical particles with mass m may be written in the form [53, 54]:

$$H = -\frac{\hbar^2}{2m}\sum_i \nabla_i^2 + \frac{4\pi a \hbar^2}{m}\sum_{i<j}\delta(\mathbf{r_i} - \mathbf{r_j}). \tag{1.81}$$

The Hamiltonian in this form is correct for both cases of Fermi and Bose statistics.

Nonideal Fermi gas

We consider a system of N interacting identical fermions with mass m and spin $1/2$ contained in the volume V at low temperatures. The interaction potential is described by the dispersion length a. For the unperturbed wave function, we take the wave function of free particles Φ_n, where n corresponds to the total occupation number $(\ldots n_p \ldots)$, n_p is the number of fermions with the momentum p and spin s. The energy level can be obtained from the relation $E_n \equiv (\Phi_n, H\Phi)$, thus we have [53]

$$E_n = \sum_p \frac{p^2}{2m}(n_p^+ + n_p^-) + \frac{4\pi a \hbar^2}{mV}N^+N^-, \tag{1.82}$$

where $n_p^\pm$ is the occupation number of particles with spin $s = \pm 1/2$ and $N^\pm = \sum_p n_p^\pm$. This expression is valid under the condition $k_F|a| \ll 1$, where k_F is the Fermi wave vector. For $a > 0$ and repulsive particle potential, the minimum of energy is attained when N^+N^- and all spins are oriented along the same direction. For $a < 0$, the minimum of energy is attained if the neighbor spins have opposite orientations. The partition function for this system may be taken in the standard form given by

$$Z_N = \sum_{n_p^\pm} \exp\left\{-\beta\left[\sum_p \frac{p^2}{2m}(n_p^+ + n_p^-) + \frac{4\pi a \hbar^2}{mV}N^+N^-\right]\right\}. \tag{1.83}$$

If it is possible to calculate the partition function, then we can describe the system. But in the general case, it is impossible.

Non-ideal Bose gas

In this section we consider a system of N Bose particles whose interaction potential is characterized by the dispersion length a. For the unperturbed wave function we take the wave function of free particles Φ_n, where n corresponds to the total occupation number $(\ldots n_p \ldots)$, n_p is the number of bosons with the momentum p and spin s. The energy level may be obtained from the relation $E_n \equiv (\Phi_n, H\Phi)$ in a manner similar to the case of Fermi particles and as a result we have [53]

$$E_n = \sum_p \frac{p^2}{2m} n_p + \frac{4\pi a \hbar^2}{mV}\left(N^2 - \frac{1}{2}\sum_p n_p^2\right), \qquad (1.84)$$

where n_p is the particle occupation number and $N = \sum_p n_p$. This expression is valid under the condition $a/v^{1/3} \ll 1$ and $ka \ll 1$. If we take $n_p = 0$ and introduce $n_0 = N$, then we obtain the energy of the ground state per particle as given by

$$\frac{E_0}{N} = \frac{2\pi a \hbar^2}{mv}, \qquad (1.85)$$

where $v = V/N$. For the excited state, the energy per particle takes the value

$$\frac{E_n}{N} = \frac{2\pi a \hbar^2}{mv}\left[1 - \frac{1}{2}\sum_p \frac{n_p^2}{N^2}\right]. \qquad (1.86)$$

The partition function for this system may be taken in the standard form

$$Z_N = \sum_{n_p} \exp\left\{-\beta\left[\sum_p \frac{p^2}{2m} n_p + \frac{4\pi a \hbar^2}{mV}(N^2 - n_0^2)\right]\right\}. \qquad (1.87)$$

In this case, we cannot calculate the partition function exactly. Nevertheless, we can write the probable equation of state in the manner similar to the case of Fermi particles.

Chapter 2

Statistical Description
of Phase Transitions

2.1 Theory of the Second-Order Phase Transitions

Bragg–Williams theory

We start the description of phase transitions from the theory of Bragg and Williams. In this model, we introduce the order parameter as the mean physical quantity that describes probable ordering in the system. In the Ising model, the order parameter is the averaged spin $m = \langle s \rangle$. The free energy of the system is the difference of the internal energy and the entropy. The entropy for a given spatially uniform m may be calculated exactly. The total magnetic moment is determined by the difference of the numbers $N_\uparrow$ of particles with spins up and $N_\downarrow$ of particles with spins down, i.e., $m = (N_\uparrow - N_\downarrow)/N$, where $N = N_\uparrow + N_\downarrow$ is the total number of sites in the lattice. The entropy of such a system may be presented as a logarithm of the number of configurations. i.e.,

$$S = N \ln 2 - \frac{N}{2} \left\{ (1 + m) \ln(1 + m) + (1 - m) \ln(1 - m) \right\}. \quad (2.1)$$

This expression describes the entropy of mixing. To obtain the energy of the system we have to calculate $\langle H \rangle = Z_m \text{Tr} \left(e^{-\beta H} H \right)$ where Tr is the trace over all configurations with fixed m and $Z_m \text{Tr} \left(e^{-\beta H} \right)$. The exact calculation in this theory would be rather complicated as in the Ising model. In the Bragg–Williams theory, $m = \langle H \rangle$ is approximated by replacing σ_i by its position-independent m, i.e.,

$$E = -I \sum_l m^2 = -\frac{1}{2} I z N m^2, \tag{2.2}$$

where z is the number of nearest-neighbor sites in the lattice ($z = 2d$ for a d-dimensional hyperbolic lattice). The complete Bragg–Williams free energy is given by

$$f(m) = \frac{E - TS}{N} = -\frac{1}{2} I z m^2 - \frac{T}{2} \{(1+m)\ln(1+m)$$

$$+ (1-m)\ln(1-m)\}. \tag{2.3}$$

In the vicinity of the critical temperature T_c, where m is small, we can expand $s(m)$ and $f(m)$ in power series, so that

$$f(m) = \frac{1}{2}(T - T_c)m^2 + \frac{1}{12} T m^4 - T \ln 2 + \cdots, \tag{2.4}$$

where $T_c = zI$ is the mean-field temperature. For $T > T_c$, the values of this function are positive for the minimum at the origin, for $T < T_c$ its values are negative at the maximum and positive at the minimum for nonzero m. The equation of state in the external field is given by

$$\frac{\partial f(m)}{\partial m} = -zIm + T \tanh^{-1} m = h. \tag{2.5}$$

The solution of this equation is given by

$$m = \tanh\left(\frac{h + T_c m}{T}\right). \tag{2.6}$$

The quantity $h_m = h + T_c m$ is the average local or molecular field at a given site. For $h = 0$, we have

$$m = \tanh\left(\frac{T_c m}{T}\right) \simeq 1 - 2 \exp\left\{-\frac{2zI}{T}\right\}$$

so that m tends to one exponentially with temperature. For $T \to T_c$, the solution may be obtained by expanding the $f(m)$ in power series

$$f(m) \simeq (T_c/T)m - (1/3)(T_c^3/T^3)m^3 \tag{2.7}$$

and $m \simeq \pm[(T_c - T)/T]^{1/2}$. Thus m tends continuously to zero as $(T_c - T)^{1/2}$. This behavior is the general feature of the second-order mean-field phase transitions. The Bragg–Williams free energy may be employed to calculate the thermodynamic quantities, i.e., the internal energy and entropy, and their derivatives.

Landau theory

In what follows we describe most of these quantities in terms of the more phenomenological Landau mean-field theory. The Landau theory describes the phase transitions in terms of the order parameter. Under the simple assumptions that the information order parameter is small and the free energy is uniform near the temperature of phase transition, the order parameter completely determines this phase transition. Equilibrium thermodynamics completely determines the free energy $F(T, \langle \varphi \rangle)$, where $\langle \varphi \rangle$ is the local order parameter. The free energy is a function of the order parameter and should be invariant with respect to the symmetry group of the disordered phase. This means that the free energy can be only a function of the order parameter $\langle \varphi \rangle$ that does not change under all the operations in this group. It is possible in almost all cases to define the order parameter so that it is spatially uniform in equilibrium in the ordered phase. The free energy may be expressed in terms of the local free energy density $f(T, \langle \varphi \rangle)$. A simple expression for the free energy is given by

$$F = \int d^d x f(T, \langle \varphi \rangle) + \frac{k}{2} \int d^d x (\nabla \langle \varphi \rangle)^2, \qquad (2.8)$$

where d is the dimension of the space and k is a phenomenological coefficient. f may be written as a power series of $\langle \varphi \rangle$, i.e.,

$$f(T, \varphi) = \frac{1}{2} a \varphi^2 - c \varphi^3 + b \varphi^4. \qquad (2.9)$$

This expression does not contain the linear term for symmetry reasons. Each term should be invariant under the operation of group symmetry. All the coefficients a, b, c may, in principle, depend on the temperature. If the system has a scalar order parameter, then $f(T, \varphi)$ should be invariant under $\langle \varphi \rangle \to -\langle \varphi \rangle$, i.e., only even powers of $\langle \varphi \rangle$

are allowed in the expansion of f. In this case, we have

$$f(T, \varphi) = \frac{1}{2}a\varphi^2 + b\varphi^4. \tag{2.10}$$

For high temperatures φ should vanish when the external field h is zero. This means that f should have a minimum for $\varphi = 0$. For low temperatures, we expect an ordered state and f to have at least one minimum with nonzero φ. The easiest way to satisfy this requirement is to allow a to change the sign for some temperature T_c, $a = \alpha(T - T_c)$. Suppose b is temperature-independent. The equation of state for uniform h is given by

$$a\varphi + 4b\varphi^3 = h. \tag{2.11}$$

If $h = 0$, then the solution is $\varphi = 0$ for $T > T_c$ and $\varphi = \pm(-a/4b)^{1/2}$ for $T < T_c$. The Landau mean-field theory predicts a second-order phase transition with $\varphi \sim (T_c - T)^\gamma$, where $\gamma = 1/2$. γ is the so-called critical index that governs the temperature-dependence of the order parameter in the vicinity of T_c. In the mean-field theory, its value is equal to $1/2$. When critical fluctuations are important, γ is in the general case smaller than its mean-field value, typically of the order of $1/3$ in three-dimensional systems. In this system, it is possible to obtain the susceptibility by differentiating the previous equation with respect to φ, we have

$$(a + 12b\varphi^2)\frac{\partial\varphi}{\partial h} = 1, \tag{2.12}$$

and thus we obtain

$$\chi = \frac{\partial\varphi}{\partial h} = \frac{1}{a + 12b\varphi^2}. \tag{2.13}$$

Its behavior is described by $\chi = 1/a$ for $T > T_c$ and $\chi = 1/2a$ for $T < T_c$. This implies that $\chi = |T - T_c|^{-\delta}$ where $\delta = 1$ and is called the susceptibility exponent. In the three-dimensional case, where critical fluctuations are important, this critical index is of the order of $4/3$. For this second-order phase transition we can obtain the jump discontinuity of the phase-transition heat

$$c_V = -T\frac{\partial^2 f}{\partial T^2} = \frac{T\alpha^2}{8b}.$$

This equation gives the specific heat associated with the ordering. The total specific heat contains a temperature term arising from the degrees of freedom not associated with the ordering. In the case when the external field is present we have different behaviors of the system. For $h \neq 0$, the equation of state has no solution $\varphi = 0$ and the order parameter has nonzero values for different V and T. The phase transition for $h \neq 0$ does not occur. If $a > 0$ and the external field is small, then $h \ll h_c = a^{3/2}b^{-1/2}$ and there exists only one solution, i.e., $\varphi = h/a$. If $a < 0$ and $h = 0$, there exist three solutions $\varphi = 0, \varphi = \pm(a/4b)$. Three solutions exist also for small values of the external field $h < h_c$. For greater external field $h > h_c$, the first term of the equation of state is small $h = b\varphi^3$, and the order parameter depends on the external field. In this theory, we can calculate the correlation function $\chi(r, r')$ using the mean-field form of the free energy including the gradient term. Thus, we have

$$\chi^{-1}(r, r') = \frac{\delta^2 F}{\delta\varphi(r)\delta\varphi(r')} = (a + 12b\varphi^2 - k\nabla^2)\delta(r - r') \qquad (2.14)$$

or, in the momentum space,

$$\chi(q) = \frac{1}{a + 12b\varphi^2 + kq^2}. \qquad (2.15)$$

The latter relation may be reduced to the form

$$\chi(q) = \frac{\chi}{k(1 + (\xi q)^2)} \equiv \frac{\xi^2}{1 + (\xi q)^2}, \qquad (2.16)$$

where

$$\xi = \frac{k^{1/2}}{(a + 12b\varphi^2)^{1/2}}$$

is the correlation length that behaves as $\xi \sim |T - T_c|^{-\nu}$, where ν is the correlation length exponent that is equal to $1/2$ in the mean-field theory and is of the order of $2/3$ in most three-dimensional critical systems. The existence of the correlation length ξ follows from a simple dimensional analysis of the model-free energy. If φ is dimensionless, then a has the dimension erg $\times$ m^{-d} and k has the dimension erg $\times$ m^{d-2}. Then $\xi \sim (k/a)^{1/2}$ should have the dimension of length. This implies that we have to introduce a temperature-independent bare correlation length $\xi_0 = (k/\alpha T_c)^{1/2}$ that determines the microscopic length scale. Near the temperature of the phase transition the correlation length $\xi \sim \xi_0 |(T - T_c)/T_c|^{-\nu}$ can be arbitrarily larger than ξ_0. The Landau theory is also valid when fluctuations occur in the volume with the linear length ξ smaller than the equilibrium value φ_0. From this condition, we obtain the criterion of applicability of the mean-field Landau theory. This criterion is given by

$$\frac{b^2 T_c^2}{\alpha k^3} \ll \frac{T - T_c}{T_c} \ll 1. \tag{2.17}$$

This parameter is constant that is specific to any condensed system, and is different for different systems.

2.2   Unification of the Theories of Phase Transitions

Within the context of the general theory of phase transitions, a system treated as a continuous medium is assumed to have a ground state that may always be described in terms of the order parameter. This order parameter may have various geometrical presentations, for example, a scalar field in the case of condensed matter [65], a fundamental scalar field in quantum field theory [78], a magnetization vector in theory of magnetism [74], a second-rank tensor in liquid-crystal theory [134], etc. To introduce the order parameter that determines a stable state of the condensed matter, we have to consider probable deformations of the field distribution, in particular, the disordered configuration of the ground state. This distribution of the order parameter can be observed experimentally. Any experiment requires that the theory should provide an appropriate explanation of the observed behavior of the order parameter that is introduced. Today, many phenomenological models exist that claim to give a general description of the phase transitions and to predict spatial distributions of the order parameters before and after phase transitions. These well-known phenomenological models are, e.g., the Landau theory of phase transitions [65] and the gradient theory of phase transitions [136, 137]. The first model gives an efficient treatment of the order parameter, while the second one describes the phase transitions in terms of the order-parameter gradient. Both models efficiently describe many details of phase transitions and the behavior of the order-parameter spatial distribution after the phase transition.

The purpose of this section is to show that the standard model of phase transitions may be unified with the gradient theory of phase transitions treated in terms of the order-parameter gradient. A generalization of the gradient theory of phase transitions with regard to the fourth power of the order parameter and its gradient is proposed. Such generalization makes it possible to describe a wide range of phase transitions within a unified approach.

In particular, it is consistent with the nonlinear models [59, 111] which describe phase transitions accompanied by spatially inhomogeneous distributions of the order parameter. We show the formation of spatially inhomogeneous distributions of the order parameter in the course of phase transitions to be a characteristic feature of many nonlinear models of phase transitions. The solutions of the standard and gradient models of phase transitions are shown to be sometimes inconsistent with the Euler–Lagrange equation generated by the generalized free-energy functional. In the general case, the consistency requirement of the known solutions with the generalized description can be satisfied by an appropriate choice of the coupling between the order parameter and its gradient. An example of the model of phase transitions with coupling between the order parameter and its gradient is considered. We obtain an exact solution for this model, predict experimental observations of the behavior of the order parameter in the case of the spinodal decomposition, and describe the behavior of the fundamental scalar field.

Specific examples considered in what follows show that the structural formation observed experimentally can be described by various phenomenological free-energy functionals associated with various sets of coefficients, i.e., we have no unique functional representation of the free energy associated with the chosen spatially inhomogeneous configuration of the order parameter. This uncertainty is generated by the phenomenological description and could be eliminated in the microscopic calculations. We present simple examples that can be consistent with the phenomenological description of phase transitions in both standard and gradient theories.

Landau model of phase transitions

We start from the standard well-known Landau model of phase transitions. The phase transition associated with the system with broken continuous symmetry may be described in terms of the relevant order parameter. In particular, according to the Landau theory, the free energy density may be presented in terms of the order parameter as

$$F = C \left(\nabla \varphi(\mathbf{r}) \right)^2 + W(\varphi(\mathbf{r})), \qquad (2.18)$$

where $\varphi(\mathbf{r})$ is the order parameter, $W(\varphi(\mathbf{r}))$ describes the order parameter dependence on the free energy that is assumed to be known, and C is a constant.

In the well-known standard model of phase transitions we have

$$W(\varphi(\mathbf{r})) = -\frac{1}{2} A \varphi^2(\mathbf{r}) + \frac{1}{4} B \varphi^4(\mathbf{r}). \qquad (2.19)$$

With the dimensionless variable $\varphi^2(\mathbf{r}) = [B\varphi^2(\mathbf{r})]/A$ being introduced, the standard dimensionless free energy depends on the relation between the constants A and B and reduces to the form given by

$$f = l^2 \left(\nabla \varphi(\mathbf{r}) \right)^2 + \left(1 - \varphi^2(\mathbf{r}) \right)^2 \qquad (2.20)$$

for $A = B$ or

$$f = l^2 \left(\nabla \varphi(\mathbf{r}) \right)^2 + \left(1 - \varphi^2(\mathbf{r}) \right) \varphi^2(\mathbf{r}) \qquad (2.21)$$

for negative A and $B = -2A$. Here, the inverse potential is written in the standard form and $l^2 = 4(C/A)$ is the square of the characteristic length. Making use of these expressions for the free-energy density, we find the spatial distribution of the order parameter and thus describe the properties of the new states that can be formed after the phase transition. In the first case, the stationary distribution of the order parameter may be obtained from the Euler–Lagrange equation

$$l^2(\nabla)^2\varphi(\mathbf{r}) = 2\varphi(\mathbf{r})(\varphi^2(\mathbf{r}) - 1) \qquad (2.22)$$

that has a topological soliton solution $\varphi = \tanh(lr)$. In the second case, we obtain another form of the Euler–Lagrange equation, i.e.,

$$l^2(\nabla)^2\varphi(\mathbf{r}) = \varphi(\mathbf{r})(1 - 2\varphi^2(\mathbf{r})), \qquad (2.23)$$

whose solution $\varphi = \mathrm{sech}\,(lr)$ does not contain the topological charge. From this example, we may conclude that different coefficients in the expansion of the free energy produce differing spatial distributions of the order parameter that can be observed experimentally.

It should be noted that in the case of a system described by the gradient of the order parameter, it looks reasonable to introduce into the free-energy functional a term responsible for probable interaction between the order parameter and its gradient. This coupling can regularize probable perturbations of the order parameter and thus confine the spatially inhomogeneous state of the system. The point is that not any deformation of the order parameter can exist in the deformed matter. Each defect requires additional energy and deformations consistent with the spatial distribution of the order parameter are not arbitrary. On the other hand, singular perturbation models, involving higher orders of the spatial derivatives, provide a new understanding of the role of additional physical features of the system considered in phase transitions, in particular, the way to describe the contribution of the surface energy. Strong deformations exist in a small spatial interval and we have to introduce higher orders of the spatial derivative. In this case, with the accuracy of the free-energy functional being restricted to the first-order derivatives of the order parameter, we obtain a solution with homogeneous distribution of the order parameter only.

General description of the gradient theory of phase transitions

The main idea of this subsection is to generalize the phenomeno-logical theory of phase transitions by introducing the second-order spatial derivatives into the order-parameter dependence of the free energy. Such generalization is reasonable in view of both mathemat-ical and physical arguments. From the mathematical point of view, our argument is that the order parameter may be treated as a vector quantity and thus we may rewrite the free energy density in the form

$$f = l^2 \left(\nabla \varphi(\mathbf{r}) \right)^2 + \left[\left(|\varphi(\mathbf{r})| \right)^2 - 1 \right]^2, \qquad (2.24)$$

where the order parameter is a vector function. This problem arises if we assume that the order parameter $\varphi(\mathbf{r})$ may be presented as a gradient of some other scalar function $\nabla u(\mathbf{r})$. This assumption leads to the known presentation of the free-energy density of the Aviles–Giga form [136, 137], i.e.,

$$f = l^2 \left(|\nabla \nabla u(\mathbf{r})| \right)^2 + \left[\left(|\nabla u(\mathbf{r})| \right)^2 - 1 \right]^2. \qquad (2.25)$$

This presentation has various physical applications, e.g., the descrip-tion of smectic liquid crystals [136], thin film blisters [138, 139], and convective pattern formation [140]. Physically, such a model may be regarded as the Landau model applied to a system with vector order parameter. Some well-known Landau theories have similar features. For example, the energy of a smectic-A liquid crystal has been described within this model [136] with $[\nabla u(\mathbf{r})]/[u(\mathbf{r})]$ representing the directing field of the liquid crystal. The observed focal-conic defect structures may also be described in terms of this functional [137]. Micromagnetics give one more example of the application of the functional (2.25). In particular, it may be used to describe the magnetization constrained by $|m| = 1$ within and $m = 0$ outside the micromagnet [117]. The unknown $u(\mathbf{r})$ is purely curl-free, while $|m|$ just prefers to be divergence-free. Moreover, $|m|$ is restricted to unit vectors, while $u(\mathbf{r})$ prefers to have magnitude one. But the similarity should be clear, especially for an isotropic ferro-magnet [65]. Another motivation to use the gradient theory of phase transitions is provided by structural phase transitions in which the order parameter is a homogeneous deformation of the crystal [141]. In these cases, the expression for the free-energy density contains the invariants of fourth-order derivatives of the displacements. This approach describes the acoustic phonon instabilities and structural phase transitions of various crystals [141].

Another motivation to use the functional (2.25) arises from recent phenomenological modeling of blisters in compressed thin films [138]. Formerly it has been suggested that the fold patterns of such blisters could be described by minimizing the sum of membrane and bending energies [65]. With some simplification, this problem may be reduced to the free-energy density given by (2.25). The same free energy may also be obtained for an equilibrium state of a free surface. In this case, $u(\mathbf{r})$ represents the height profile of the sheet (relative to the flat reference state). This model describes a fluctuating fluid membrane.

The interpretation of the free energy appears also in the phase diffusion theory of pattern formation proposed by Cross and Newell [142, 143]. Of course, physical justification of the models discussed above may be criticized [144]. Nevertheless, they demonstrate the need to introduce new representations for the free-energy functional in terms of the order parameter. We have referred informally to the existence of an asymptotic variation problem. Now we consider the possibility to generalize the representation of the free energy to the case when it depends not only on the order parameter and its gradient, but also on their combination. Such generalization looks quite reasonable if one bears in mind that, in the case of a functional with the first-order spatial derivative, the order parameter describes a phase transition in a spatially homogeneous system, while the presence of the second-order derivative makes it possible to describe the formation of spatially inhomogeneous structures. Thus, we may expect that the combination of the gradient terms with the scalar order parameter can be responsible for the self-consistent influence of the order parameter on the spatial distribution and parameters of the ordered structures. In other words, the coupling between the order parameter and its derivatives can influence probable deformations and confine probable inhomogeneous stable structures of the system.

So, we assume that the order-parameter dependence on the free energy is given by

$$f = al^2(|\nabla\varphi(\mathbf{r})|)^2 + b((|\varphi(\mathbf{r})|)^2 - 1)^2 + ml^4(|\nabla\nabla\varphi(\mathbf{r})|)^2$$
$$+ n(l^2(|\nabla\varphi(\mathbf{r})|)^2 - 1)^2 + c\varphi^2(\mathbf{r})(l\nabla\varphi(\mathbf{r}))^2, \tag{2.26}$$

where l is the length of the order-parameter variation, a, b, m, n, and c are parameters describing the influence of the order-parameter gradient and the coupling between the order parameter and the relevant derivative of this parameter. Having introduced the operator $D = l\nabla$ we can rewrite the free-energy density as

$$f = a(|D\varphi(\mathbf{r})|)^2 + b((|\varphi(\mathbf{r})|)^2 - 1)^2 + m(|D^2\varphi(\mathbf{r})|)^2$$
$$+ n(l^2(|D\varphi(\mathbf{r})|)^2 - 1)^2 + c\varphi^2(\mathbf{r})(D\varphi(\mathbf{r}))^2. \tag{2.27}$$

If all the coefficients except a and b are equal to zero, then we come to the standard theory of phase transitions. If a, b, and c vanish, then we come to the standard gradient theory. In the case of large values of m (m is larger than other coefficients) we obtain the eikonal equation $\nabla\varphi(\mathbf{r}) = 1$ and supplement it with the condition $\varphi(\mathbf{r}) = 0$ at the boundary. This eikonal equation has no smooth solution, but it has infinitely many Lipschitz solutions. We may suggest that the energy can be concentrated at the discontinuities of $\varphi(\mathbf{r})$. Thus, the singular part of the order parameter can provide a selection mechanism for the perturbations of the eikonal equation solutions that minimize the free energy.

In the general case, the minimum of the free energy satisfies the Euler–Lagrange equation

$$\frac{\delta f}{\delta\varphi(\mathbf{r})} = \frac{\partial f}{\partial\varphi(\mathbf{r})} - D\frac{\partial f}{\partial D\varphi(\mathbf{r})} + D^2\frac{\partial f}{\partial D^2\varphi(\mathbf{r})} \tag{2.28}$$

that reduces in our case to

$$mD^4\varphi(\mathbf{r}) - (a - 2n + 6n(D\varphi(\mathbf{r}))^2 + c\varphi^2(\mathbf{r}))D^2\varphi(\mathbf{r})$$
$$- c\varphi(\mathbf{r})(D\varphi(\mathbf{r}))^2 - 2b\varphi(\mathbf{r})(1 - \varphi^2(\mathbf{r})) = 0. \tag{2.29}$$

Probable solutions of the Euler–Lagrange equation

In this section, we analyze probable solutions of the Euler–Lagrange equation for various combinations of coefficients in the one-dimensional case.

(a) Linear solutions:

(i) We assume that the solution is similar to the solution of the equations $D\varphi = -\varphi$ and $D^2\varphi = \varphi$. Substituting this solution in the Euler–Lagrange equation yields

$$m\varphi - (a - 2n + c\varphi^2 + 6n\varphi^2)\varphi - c\varphi^3 - 2b\varphi(1 - \varphi^2) = 0.$$

$$(2.30)$$

As follows from the latter equation, the coefficients satisfy the relations $m - a = 2(b - n)$ and $b - c = 3m$. In the standard model of phase transitions $a = b = 1$ and thus we have $m + 2n = 3$ and $c = 1 - 3n$. If $m = n = 1$, then for the coupling constant, we have $c = -2$. If $a = 1$, $b = -1$, and $m = 1$, $n = -1$, then the coupling constant $c = 2$. In the case of the gradient presentation $a = b = 0$, the exponential solution can be realized only for $n = 1$, $m = -2$, and $c = 3$. For the standard model with $n = m = 0$, the exponential solution exists for $b = 1$, $a = -2$, and $c = 1$. Thus, the exponential solution is realized for various combinations of coefficients.

(ii) Now let us seek for another probable linear solution in the form of a periodic function, i.e., $\varphi(\mathbf{r}) = \varphi \exp(ikr)$. This solution satisfies the equation $D^2\varphi(\mathbf{r}) = -k^2\varphi(\mathbf{r})$. Substituting it in the Euler–Lagrange equation leads to the relation

$$mk^4\varphi(\mathbf{r}) + (a - 2n + c\varphi^2 + 6nk^2\varphi^2)k^2\varphi(\mathbf{r})$$
$$- c\varphi^2\varphi(\mathbf{r}) - 2b\varphi(\mathbf{r})(1 - \varphi^2) = 0 \qquad (2.31)$$

that generates two conditions, i.e., $mk^4 + (a - 2n)k^2 - 2b$ and $6nk^2 + 2b = 0$, for which the spatial periodic distribution of the order parameter does not depend on the coupling constant. The conditions thus obtained may be employed to find the wavelength of the periodic distribution of the order parameter. The second condition yields $k^2 = -b/3n$. On the other hand, from the first condition we have $k^2 = -(a+4n)/m$. This means that these conditions are consistent for $mb = 3n(a+4n)$. For $k^2 = 1$, when the period of the structure is equal to the size of the system under consideration, we obtain $m = -(a+4n)$ or $b = -3n$. In the case of the standard theory of phase transitions with $a = b = 1$, we find the other coefficients, i.e., $n = -1/3$ and $m = 1/3$. It is important that in the case under consideration the solution for the order parameter is independent of the coupling constant. The next step is to consider the realization of probable nonlinear solutions that satisfy the minimum of the free-energy density.

(b) Nonlinear solutions:

(i) We start from the nonlinear solution of the standard model of phase transitions, i.e., from the well-known soliton solution that satisfies the equation

$$D^2\varphi = 2\varphi(\varphi^2 - 1). \tag{2.32}$$

The solution of the latter equation is $\varphi = \tanh(lr)$. It satisfies an equation with high-order derivatives that is given by $D\varphi = (1 - \varphi^2)$, and the relation $D^4\varphi = 8\varphi(1 - \varphi^2)(3\varphi^2 - 1)$. Having substituted the latter relations into the Euler–Lagrange equation, we find that this solution can be realized under the conditions $n = 0$, $16m + 2(a-b) - c = 0$ and $c = 8m$. In the case of the standard theory of phase transitions, $a = b$, all the other coefficients should be equal to zero. In the case of the generalized model, we observe the soliton solution for $a = 1$, $b = 2$, $c = 2$ and $m = 1/4$. We may also consider other nonlinear solutions of the standard model.

(ii) The second example of the nonlinear solution of the standard model is $\varphi = \operatorname{sech}(lr)$ that satisfies the nonlinear equation

$$D^2\varphi = \varphi(1 - 2\varphi^2). \tag{2.33}$$

Assuming also that both equations $D\varphi^2 = \varphi^2(1 - \varphi^2)$ and $D^4\varphi = \varphi(1 - 20\varphi^2 + 24\varphi^4)$ are satisfied, we obtain an equation for the coefficients of the generalized model. In particular, the solution in the proposed form exists for $m = a + 2b$, $c = -8m$, and $a - c + b = 10m$. This means that in the case of the standard model $a = b$ does not lead to the minimization of the generalized free-energy functional.

(iii) Finally, let us consider whether the solution $\varphi = \ln \cosh(lr)$ can be realized. This solution satisfies the equations $D^2\varphi = (1 - (D\varphi^2)$, $D\varphi = \tanh(lr)$, and $D^4\varphi = 2(1 - (D\varphi)^2)(3(D\varphi)^2 - 1)$. As follows from the Euler–Lagrange equation, this solution minimizes the free-energy functional under the conditions $a = b = c = 0$, i.e., we obtain the well-known form

$$f = (D^2\varphi)^2 + (1 - (D\varphi)^2). \tag{2.34}$$

The absolute minimum of the free energy is attained for $\varphi = 1$ and $D\varphi = 1$, that cannot be satisfied. This means that the eikonal solution satisfies the condition of the free-energy divergence. Such conditions can be realized only for spatially inhomogeneous order parameters.

Thus, we have proposed a generalization of the standard and gradient theories of the phase transitions by introducing the coupling between the order parameter and its gradient. Such generalization may be employed to describe the phase transitions from spatially homogeneous to inhomogeneous states. It is shown that the solutions of the standard and gradient models of phase transitions can be inconsistent with the Euler–Lagrange equation generated by the generalized functional of the free energy. In the general case, the requirement of consistency of the known solutions with the generalized description can be satisfied by an appropriate choice of the coupling between the order parameter and its gradient.

Specific examples considered in this section show that the structure formation observed experimentally may be described by various phenomenological free-energy functionals that correspond to various sets of coefficients, i.e., there is no unique functional representation of the free energy associated with the chosen spatially inhomogeneous configuration of the order parameter. This uncertainty could be due to the phenomenological description and might be eliminated in microscopic calculations.

Exact solution of the model of phase transitions with coupling

The main idea of this section is to show that the standard model of phase transitions and the gradient theory may be unified by considering the coupling between the order parameter and its gradient. Introducing such coupling into the free-energy density dependence makes it possible to find a spatially inhomogeneous solution of the order parameter. Today, we see several physical applications for the generalization introduced, including a spinodal decomposition [145] and an alternative cosmological model that takes into account probable formation of an inhomogeneous distribution of the fundamental scalar field [78]. As shown earlier [145], if one uses the expansion of the thermodynamic potentials with the accuracy up to the second order of parameter and its gradient, and the coefficients can change their signs, then various decomposition scenarios become probable that can result in the transitions from the disordered state to the modulated, patterned, or ordered-patterned states. In the cosmological scenario, for example, the formation of spatially inhomogeneous states may be described if we assume that coefficients in the expansion of the model potential possess some specific properties.

If the system described by a scalar space-dependent order parameter is rapidly quenching from a homogeneous high-temperature phase to the domain of two-phase coexistence, then we have the spinodal decomposition in the system. The decrease of the temperature can be a reason for the formation of a new bubble phase in the cosmological model. Usually, the theoretical analysis of the spinodal decomposition and spatial formation of the fundamental scalar field is based on the standard model. But in this model the presentation of the free energy contains no terms responsible for the interaction between the order parameter and its gradient that could introduce the restriction for the deformation of the order parameter. At the same time, any deformation of the order parameter requires additional energy and thus probable deformations that can be realized are not arbitrary. In the general case, the requirement of consistency of the known solutions with the generalized description may be achieved by an appropriate choice of the coupling between the order parameter and its gradient. In order to illustrate the efficiency of the proposed idea, we restrict the consideration to a simplified version of the general unified model, namely,

$$f = a\big(|D\varphi(\mathbf{r})|\big)^2 + b\big(|\varphi(\mathbf{r})|\big)^2 - 1\big)^2 + c\varphi^2(\mathbf{r})\big(D\varphi(\mathbf{r})\big)^2, \qquad (2.35)$$

where operator $D = l\nabla$ with the length of variation of the order parameter l is introduced, a, b, and c are parameters describing the influence of the order-parameter gradient, the order parameter, and the coupling between the order parameter and order-parameter gradient. If the coefficient c is equal to zero, then we return to the standard theory of phase transitions, or the theory of fundamental scalar field. Nevertheless, these considerations demonstrate the need to treat the variation problems for the functional in another representation. We prove that, as the relaxation parameter vanishes, the families of such fields with finite free energies are compact and have many solutions that describe spatially inhomogeneous distributions of the order parameter and probable topological structure of the new phase. Our proof is based on the analysis of the extremum of the free-energy density. The method also yields an Euler–Lagrange equation for the limit of the defect structure measure. This measure also satisfies the cancellation properties depending on its local regularity that seems to indicate several levels of singularities in the limit. We show that the same solution can be obtained in another interpretation. To do this, we have to analyze the behavior of the free-energy density. The minimum of the free energy satisfies the Euler–Lagrange equation, i.e.,

$$(a + c\varphi^2(\mathbf{r}))D^2\varphi(\mathbf{r}) + c\varphi(\mathbf{r})(D\varphi(\mathbf{r}))^2 + 2b\varphi(\mathbf{r})(1 - \varphi^2(\mathbf{r})) = 0.$$
$$(2.36)$$

Let us consider several probable solutions of the Euler–Lagrange equation for various combinations of coefficients in the one-dimensional case when $D\varphi = d\varphi(\mathbf{r})/dx = \dot{\varphi}$. The Euler–Lagrange equation may be rewritten in the simple form

$$\ddot{\varphi} + g(\varphi)(\dot{\varphi})^2 + h(\varphi) = 0, \qquad (2.37)$$

where

$$g(\varphi) = \frac{c\varphi}{a + c\varphi^2}, \quad h(\varphi) = \frac{2b\varphi(1 - \varphi^2)}{a + c\varphi^2}.$$

We introduce a new variable $u(\varphi) = (\dot{\varphi})^2$, then the equation thus obtained may be reduced to the Bernoulli equation, i.e.,

$$u'(\varphi) + 2g(\varphi)u(\varphi) + h(\varphi) = 0, \qquad (2.38)$$

where $u'(\varphi)$ implies the derivation with respect to φ. In order to prove this equation, one should make use of the relation $\dot{u}(\varphi) = u'(\varphi)u(\varphi)$. The general solution of the Bernoulli equation in our case is given by

$$u(\varphi) = \frac{1}{a + c\varphi^2}\left[s + b(1 - \varphi^2)^2\right], \qquad (2.39)$$

where s is the integration constant that in the general case may be taken equal to zero. This solution of the Bernoulli equation yields a relation that describes the spatial behavior of the order parameter, i.e.,

$$\frac{d\varphi}{dx} = \frac{b^{1/2}(1 - \varphi^2)}{(a + c\varphi^2)^{1/2}}, \qquad (2.40)$$

and thus the spatial distribution of the order parameter may be presented in the form

$$x = \left(\frac{c}{b}\right)^{1/2} \text{Arsinh}\left[\left(\frac{c}{a}\right)^{1/2}\varphi\right] + \left(\frac{a + c}{b}\right)^{1/2}\text{Arsinh}\left[\left(\frac{a}{c}\right)^{1/2}\right]$$
$$+ \frac{1}{2}\left(\frac{a + c}{b}\right)^{1/2}\text{Arsinh}\left[\left(\frac{2\varphi\sqrt{(a + c)(a + c\varphi^2)}}{a(\varphi^2 - 1)}\right)\right]. \qquad (2.41)$$

The behavior of the solutions for different signs of the coupling constant is shown in Fig. 2.1.

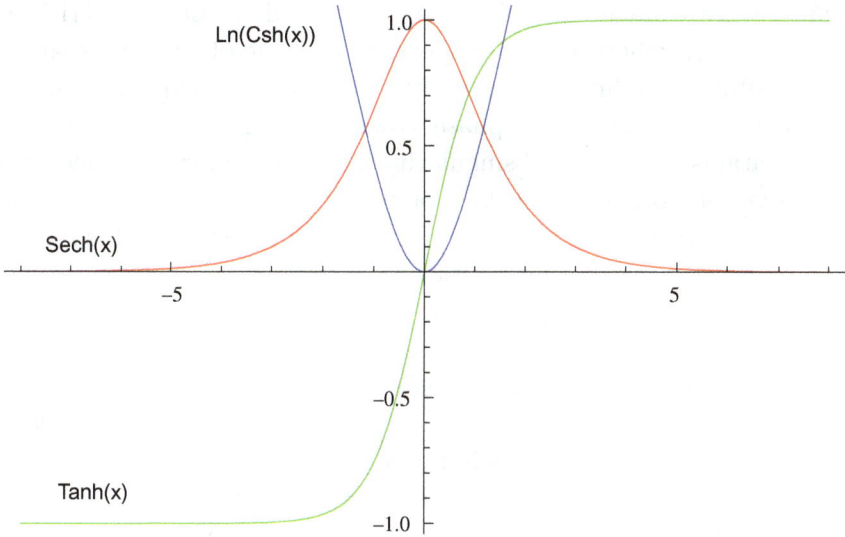

Fig. 2.1.   The exact solution for different signs of coupling constant.

It is not difficult to see that this solution yields all the well-known standard solutions. For $c = 0$, we obtain

$$\varphi = \tanh \left(\frac{a}{b} \right)^{1/2} x, \tag{2.42}$$

that completely reproduces the solution of the standard theory of phase transitions or the behavior of the fundamental scalar field in the standard cosmological model. A more interesting case occurs when $c = -a$. In this case, we obtain a simple solution $\varphi = \sin(b/a)^{1/2}x$. This solution is, however, not trivial in the physical sense. It suggests a new scenario of phase transitions. Namely, we can predict the occurrence of various spatial distributions of the order parameter after the formation of the new phase. In order to make such a prediction, we have to know (or postulate) the temperature dependence of the coupling constant. If the coupling between the order parameter and its gradient is included into the free energy with positive value, then we observe an inhomogeneous distribution of the order parameter or the fundamental scalar field without topological singularities. If the coupling constant changes as temperature changes and if its temperature-dependence is critical, $c = c(T - T_s)$, where T_s is the new critical value of the temperature for vanishing coupling constant, then we obtain an explicit solution of the standard model of phase transitions, $\varphi = \tanh{(a/b)^{1/2}x}$, that contains a topological singularity. Further temperature decrease can give rise to negative values of the coupling constant with the dependence $c = c(T - T_s) = -a$ and the topological solution transforms into a periodic solution $\varphi = \sin{(b/a)^{1/2}x}$, i.e., a regular solution without any singularities.

Assuming that the interaction between the order parameter and its gradient induces the probable deformations that depend on the coupling constant, we can obtain various spatial distributions of the order parameter. Such phase transitions can be observed in experiments with decreasing temperature. This effect may be anticipated in the standard cosmological model, when we observe the evolution of bubble formation (as a topological object) to a new periodic phase of the fundamental scalar field without singularities. This effect should be observed in experimental studies of the spinodal decomposition.

The problem of unification of the standard theory of phase transitions and the gradient theory of phase transitions may be solved by introducing coupling between the order parameter and its gradient. Such unification is needed to describe spatially inhomogeneous, in particular periodic, structure formation. The coupling can regularize the order parameter deformations and stabilize the spatially inhomogeneous state of the system. The generalization proposed here makes it possible to recover all known solutions of the standard and gradient theories by an appropriate choice of the coefficients in the approximation of the free-energy density functional of the generalized model.

The new model of phase transitions with coupling is proposed and solved. It is shown that if the coupling between the order parameter and its gradient is included in the free energy with a negative coefficient, then we observe a periodic distribution of the order parameter or the fundamental scalar field. We suggest that this effect should be observed experimentally as temperature decreases.

This conclusion can be very important in the standard cosmological model, where we have the change of the bubble structure to a periodic distribution of the fundamental scalar field.

2.3 First-Order Phase Transitions

Let us discuss the known results [65, 66, 70, 73] of the mathematical model of the first-order phase transitions in which the partition function is represented by a functional integral with the initial Gaussian distribution of occupation numbers given by

$$P(n_i) = \frac{1}{\sqrt{2\pi}} \exp\left\{-\frac{1}{2}n_i^2\right\}. \tag{2.43}$$

This approach is conceptually similar to the earlier methods, however, now n_i is a continuous variable with the distribution (2.43). After all appropriate quantities have been averaged at the more informative microscopic level, the Hamiltonian of the system may be reduced to the form [66]

$$H = -\frac{1}{2}\sum_{i,j} V_{ij} n_i n_j + A\sum_i n_i^4 - h\sum_i n_i, \tag{2.44}$$

where V_{ij} is the pair interaction energy, A is the coefficient responsible for the nonlinearity of the system, and h is the external field. Then the partition function of the system may be written in terms of a functional integral over occupation numbers, i.e.,

$$Z = \prod_{i=1}^{N} \int_{-\infty}^{\infty} \frac{dn_i}{\sqrt{2\pi}} \exp\left\{-\frac{1}{2}h\sum_{i,j}(\delta_{ij} - \beta V_{ij}) n_i n_j\right.$$
$$\left. - \beta A\sum_i n_i^4 + \beta h\sum_i n_i\right\}, \tag{2.45}$$

where

$$\beta_0^{-1} \equiv \int d\rho V(\rho), \quad B^2 = \frac{\beta_0}{3}\int d\rho \rho^2 V(\rho), \quad \rho = \mathbf{r} - \mathbf{r}'.$$

Having introduced a new variable

$$\varphi(\rho) \equiv B \left(\frac{\beta}{\beta_0} \right)^{1/2} n(\mathbf{r}),$$

we can rewrite the partition function (2.45) in terms of the latter and thus obtain the standard expression

$$Z = \exp \left\{ -\frac{N}{2} \ln \left(\frac{\beta B^2}{\beta_0} \right) \right\} \int_{-\infty}^{\infty} D\varphi \exp \left\{ -F(\varphi) \right\} \qquad (2.46)$$

with the explicit representation

$$F(\varphi) = \int d\mathbf{r} \left\{ \frac{1}{2} (\nabla \varphi)^2 + V(\varphi) \right\} \qquad (2.47)$$

and the potential energy given by

$$V(\varphi) = -\frac{1}{2} \mu^2 \varphi^2 + \frac{1}{4} \lambda \varphi^4 - \xi \varphi, \qquad (2.48)$$

where

$$\frac{1}{2} \mu^2 \equiv \frac{1}{2B^2} \left(1 - \frac{\beta}{\beta_0} \right), \quad \frac{1}{4} \lambda \equiv \frac{\beta_0^2 A}{\beta B^4}, \quad \xi \equiv \frac{(\beta \beta_0)^{1/2} h}{B}.$$
$$\qquad (2.49)$$

Thus, the description in terms of the new variable reduces to the previous procedure, however, at the level of the distribution function that serves for the order parameter, without studying the microscopic reasons that give rise to the nonlinearity. At this level, we investigate the formation of the new phase nucleus and find the determinative factor of its description. The most important contribution in the partition function is associated with the field configuration for which the value of free energy (2.47) is minimum, i.e.,

$$\frac{\delta F}{\delta \varphi} = \Delta \varphi - \frac{dV}{d\varphi} = 0. \qquad (2.50)$$

Substituting the solution of Eq. (2.50) in the expression for free energy yields the free energy and its variation due to the formation of new phase macrobubbles. The standard definition [78] of the probability of the new phase bubble formation regarding zero modes associated with an arbitrary distribution of bubble centers is given by

$$P \sim \theta^4 \left(\frac{F}{2\pi\theta} \right)^{3/2} \exp(-F). \qquad (2.51)$$

The main problem arising in the derivation of the expression for the new phase bubble size is to find the solutions of Eq. (2.50) and to derive an expression for the free energy of these configurations. Further results can be obtained without specifying the form of the potential $V(\varphi)$, it is sufficient to assume that its minima differ by some value ε. The two limiting cases exist, for which an analytic solution can be found. The first case corresponds to the situation when the difference of minimum effective potential values is greater than the barrier height, and the second case is opposite and reduces to the condition that the bubble size is much greater than its boundaries. For the sake of definiteness we restrict the consideration to the latter case, then Eq. (2.50), written in terms of the spherical coordinate system, reduces to [78]

$$\frac{d^2\varphi}{dr^2} + \frac{2}{r}\frac{d\varphi}{dr} - \frac{dV}{d\varphi} \simeq \frac{d^2\varphi}{dr^2} - \frac{dV}{d\varphi} = 0, \qquad (2.52)$$

with the limiting condition $\varphi \to 0$ for $r \to \infty$ and $\frac{d\varphi}{dr} = 0$ for $r = 0$. In the limiting case $\varepsilon \to 0$, it is equivalent to $\xi \to 0$, the equation has a solution given by

$$r = \int_\varphi^{\varphi_0} \frac{d\varphi}{\sqrt{2V(\varphi)}},$$

and the free energy is described by the expression

$$F = 4\pi \int_0^\infty r^2 dr \left\{ \frac{1}{2}\left(\frac{d\varphi}{dr}\right)^2 + V(\varphi) \right\} = -\frac{4\pi}{3}r^3\varepsilon + 4\pi r^2\sigma,$$

$$(2.53)$$

where σ is the surface energy of the bubble boundary that is equal to the action corresponding to the solution of the one-dimensional problem, i.e.,

$$\sigma = \int_0^\infty dr \left\{ \frac{1}{2}\left(\frac{d\varphi}{dr}\right)^2 + V(\varphi) \right\} = \int_0^\infty d\varphi \sqrt{2V(\varphi)}, \qquad (2.54)$$

where the integral should be calculated for $\varepsilon \to 0$. The radius of the new phase bubble is determined by the minimum of free energy; it is given by

$$r_0 = \frac{2\sigma}{\varepsilon}, \qquad (2.55)$$

whence we find that $F = (16\pi\sigma^3)/(3\varepsilon^2)$. In this case, the expression for the probability of formation of the second phase bubble reproduces the well-known result [65] often given in lecture courses. The only, though very important, difference is that the surface tension is described by a closed expression in terms of the potential energy $V(\varphi)$ and the value of minimal nonequivalence for this very potential. Thus it is clear that, for the new phase bubble to be formed, this nonequivalence should be finite, which observation suggests the appropriate form of the potential. In the case of potential energy (2.48), we have $\varepsilon = (2\mu\xi)/\lambda^{1/2}$ and $\sigma = \mu^3/3\lambda$, then $r_0 = \mu^2/(3\lambda^{1/2}\xi)$ and the effective value of the free energy variation is given by

$$\Delta F = S_{\text{eff}} = \frac{4\pi\sigma^3\lambda}{3\mu^2\xi} = \frac{4}{3}\pi\sigma r_0^2. \qquad (2.56)$$

The most suitable and often employed potential that describes the first-order phase transitions is given by

$$V(\varphi) = \frac{\mu^2}{2}\varphi^2 - \frac{\delta}{3}\varphi^3 + \frac{\lambda}{2}\varphi^4. \qquad (2.57)$$

All the required transition parameters, e.g., the new phase bubble size, the probability of its formation etc., may be expressed as combinations of the parameters introduced. In this sense, the physical picture becomes closed since all characteristics of the phase transition are determined by the nonlinearity irrespective of its formation mechanisms at the microscopic level. Attempts have been made [78] to obtain nonlinearities of this type both associated with the fluctuations of the medium and produced by the interaction with the fields of different nature. This concerns the description of phase transitions in condensed media, e.g., liquid crystals, superconductors, etc. [57, 67, 79]. In order to explain the fact that the transition in such systems is the first-order one, the mechanism has been reduced to the interaction of the scalar order parameter with the vector field that contains information on probable disclinations in the system. In terms of the mathematical procedure, the allowance for all probable configurations of such fields reduces to an additional integration, i.e.,

$$\int D\varphi \int D\mathbf{A} \exp\left\{-\int d\mathbf{r}\left[V(\varphi) + \gamma\,|(\mathbf{x} - iq\mathbf{A})\,\varphi|^2 + \frac{1}{8\pi}F_{\mu\nu}^2\right]\right\},$$

(2.58)

where

$$F_{\mu\nu} = \frac{\partial A_\mu}{\partial x_\nu} - \frac{\partial A_\nu}{\partial x_\mu}$$

is the stress tensor. If the field varies much faster than the order parameter, then the continual integration over such fields yields [57]

$$Z \sim \int D\varphi \exp\left\{-\int d\mathbf{r}\left[\frac{1}{2}(\nabla\varphi)^2 + \frac{1}{2}\mu^2\varphi^2 - \frac{\delta}{3}\varphi^3 + \frac{1}{4}\lambda\varphi 4\right]\right\},$$

(2.59)

where $\delta = 3\gamma q^2 (32\pi\gamma q)^{1/2}$ is determined by the interaction between the order parameter and the fluctuation vector field. We have already mentioned that such fields describe the behavioral specifics of the system associated with the fluctuation-produced disclinations in liquid crystals, eddies in semiconductors, or monopoles in vacuum [57, 78, 146]. The allowance for such fluctuations results in the change of the transition order since the effective potential for the order parameter takes the form of an expression equivalent to (2.47) with the explicit value of the parameter δ. The probable fluctuations occurring in the system may be also described in terms of the scalar field. To do this, it is sufficient to assume the fluctuating field to be the field ξ that is external with respect to the order parameter and to carry out averaging over this field. If the values of this field are distributed according to the Gaussian law, then we may write

$$Z \sim \int D\varphi \int D\xi \exp \left\{ - \int d\mathbf{r} \left(\frac{1}{2} (\nabla \varphi)^2 - \frac{1}{2} \mu^2 \varphi^2 \right. \right.$$
$$\left. \left. + \frac{1}{4} \lambda \varphi^4 - \xi \varphi \right) - \frac{\xi^2}{2\Delta^2} \right\}, \tag{2.60}$$

where Δ^2 is dispersion. The result would not also change in principle under the assumption of existing spatial-coordinate dependence of the field fluctuations ξ. Physically this picture corresponds to the systems with several degrees of freedom for which special order parameters exist. The systems of this type are glass, substitution and interstitial solid solutions, as well as unordered materials. In our special case (2.60), the fluctuation field might be associated with the temperature or local variations of the microscopic interaction energy. Integration over all probable configurations of the external field that varies faster than the order parameter does not lead to any changes of the order of phase transition. Integrating the Gaussian integral over ξ produces only renormalization $\mu^2 \rightarrow \mu^2 - A^2$ and can cause just a shift of the phase transition temperature. Such temperature shift has been observed under the noise-induced transitions [76]. In case we assume that the external field fluctuations are slower than order parameter variations, we have to first carry out averaging over the faster field, i.e., the order parameter for the probable formation of a new phase bubble for fixed ξ. Then, within the context of (2.60), we may write

$$Z \sim \int D\xi \exp \left\{ - \left(\frac{4\pi\sigma 3\lambda}{3\mu^2 \xi^2} + \frac{1}{2} \frac{\xi^2}{\Delta^2} \right) \right\}. \tag{2.61}$$

Now we can average over the slow field ξ. To do this, we have to find its condensate value from the minimum of the exponent in (2.61). Thus we obtain

$$\xi^2 = \left(\frac{8\pi\sigma^3 \lambda}{3\mu^2 \Delta^2} \right)^{1/2}, \tag{2.62}$$

hence the mean size of the new phase bubble is given by

$$r_c = \frac{2\sigma}{\varepsilon} = \frac{1}{2}\left(\frac{2\mu}{\pi\Delta}\right)^{1/4},$$ (2.63)

whereas the probability of the new phase macrobubble formation is described by the standard expression

$$P \sim \exp\left\{-\frac{4\pi}{3}\sigma r_c^2\right\}.$$ (2.64)

Thus the "condensing" value of fluctuations of the field that is "external" with respect to the order parameter completely determines both the mean critical size of the new phase bubble and the probability of its formation. The characteristics of such formations contain no free parameters other than fluctuation dispersion. In this sense, the approach is similar to the above microscopic description of the cluster formation in condensed media. Now we still have to consider the processes occurring under the relaxation of metastable states accompanied by the formation of thermodynamically stable bubbles of the new phase.

Debye–Huckel theory

In this section, we consider the application of the variation mean-field theory to classical equilibrium plasmas with moving charges of unlike signs. This theory provides a very good description of electrolytes with various charged ions in a solution and is also used to describe the disordered phase of a system containing topological point defects with Coulomb interaction. For the sake of simplicity, we restrict the consideration to a plasma with only two types of charge carriers, with positive charge $+q$ and negative charge $-q$. The overall charge neutrality requires that $N^+q^+ + N^-q^- = 0$, where $N^{\pm}$ are the numbers of particles with positive and negative charges. We denote the positions of positive and negative charges by $x_{\alpha\beta}^{\pm}$, then the Hamiltonian in an external electric field is given by

$$H = \frac{1}{2}\sum_{\alpha,\beta,\sigma,\sigma'} q^\sigma q^{\sigma'} U(x_\alpha^\sigma - x_\beta^{\sigma'}) + \sum_{\sigma,\sigma'} q^\sigma \phi^{\text{ext}}(x_\alpha^\sigma),$$ (2.65)

where $U(x) = x^{-1}$ is the Coulomb potential in the three-dimensional case. We would like to find the charge and charge-density response functions in the mean-field theory. As the first step, we write the total density matrix ρ in terms of the one-particle density matrices $\rho^{\pm}(x_i^\sigma)$ for positive and negative charges, i.e., $\rho = \prod_\sigma \rho^+(x_+^\sigma) \prod_{\sigma'} \rho^-(x_-^\sigma)$. The average numbers of positive and negative charges are proportional to the one-particle density matrices $n^{\pm} = N^{\pm}\rho^{\pm}$. The total charge density is $\rho_q = q^+ n^+(x) + q^- n^-(x)$ and the free energy may be written in the form

$$F_{\text{int}} = \frac{1}{2}\int d^3x \rho(x) U(x - x')\rho(x') + \int d^3x \rho(x)\phi^{\text{ext}}(x) \qquad (2.66)$$

with an additional term associated with the entropy, i.e.,

$$F_{\text{entropy}} = T\int d^3x n^+ \ln \frac{n^+}{N^+} + T\int d^3x n^- \ln \frac{n^-}{N^-}. \qquad (2.67)$$

The mean-field equation of state for the average density $n^{\pm}(x)$ is obtained by minimizing $F = F_{\text{int}} + F_{\text{entropy}}$ subject to the constraint that the total number of particles is fixed, i.e.,

$$\frac{\delta F}{\delta n^{\pm}}(x) = T\left(\ln \frac{n^{\pm}(x)}{N^{\pm}} + 1\right) + q^{\pm}\phi = \mu^{\pm}, \qquad (2.68)$$

where $\mu^{\pm}$ are the chemical potentials and

$$\phi(x) = \int d^3x U(x - x')\rho(x') + \phi^{\text{ext}} = \phi^{\text{ind}} + \phi^{\text{ext}} \qquad (2.69)$$

is the electric potential arising from the induced charge. When the chemical potentials are chosen so that $\int d^3x n^{\pm} = N^{\pm}$, the densities of positive and negative charges satisfy the relations

$$n^{\pm}(x) = \frac{N^{\pm}}{Z^{\pm}} \exp\left(-\frac{q^{\pm}\phi(x)}{T}\right), \qquad (2.70)$$

where

$$Z^{\pm} = \int d^3x \exp\left(-\frac{q^{\pm}\phi(x)}{T}\right). \qquad (2.71)$$

Given that the total potential $\phi(x)$ depends on the induced charge density, these equations should be solved self-consistently with respect to the density $n^\pm(x)$. For $\phi^{\text{ext}} = 0$, the self-consistent solution is trivial: the charge density and the potential are equal to zero and $n^\pm = N^\pm/V$, where V is the volume of the system. Making use of the mean-field equation we can calculate the charge density response function and the dielectric constant of the classical plasma. The charge-density response function is given by

$$\chi_\rho = -\frac{\delta\rho_q}{\delta\phi^{\text{ext}}}. \tag{2.72}$$

The sign of this relation depends on the coupling with the external field. The charge-density response function may be reduced to

$$\chi_\rho = -\frac{k^2}{4\pi}\frac{\delta\phi}{\delta\phi^{\text{ext}}}, \tag{2.73}$$

where one should bear in mind that

$$\frac{q^+(n^+)^2}{N^+} + \frac{q^-(n^-)^2}{N^-} = V^{-2}(q^+N^+ + q^-N^-) = 0 \tag{2.74}$$

and

$$k^2 = \frac{4\pi}{T}\left[(q^+)^2 N^+ + (q^-)^2 N^-\right] \tag{2.75}$$

is the square of the reciprocal Debye–Huckel screening length. Thus we see that the charge-density response function is proportional to the total potential for which we obtain

$$\frac{\delta\phi}{\delta\phi^{\text{ext}}} = \delta(x - x') - \int d^3y\, U(x - y)\chi_{\rho\rho}(y, x'). \tag{2.76}$$

The self-consistent equation for the charge-density response function may be solved using the interaction potential in the Fourier presentation, i.e., $U(x) = 4\pi/q^2$. Thus we obtain

$$\chi_{\rho\rho} = \frac{k^2}{4\pi}\frac{q^2}{q^2 + k^2}. \tag{2.77}$$

The charge-density response function tends to zero as q tends to zero, indicating that the charge density is incompressible for $q = 0$. This results from the long-range nature of the Coulomb potential. The mean-field theory provides a powerful yet relatively simple description of Coulomb gases and it is interesting to know when it is breaking down. In the mean-field theory, individual position coordinates are replaced by the average density that is spatially uniform in the absence of external fields. It thus ignores altogether the probable presence of boundary conditions. It makes a good approximation if the kinetic energy is greater than the potential energy. The mean-field theory is valid for $Tn^{1/3} > q^2$ that is possible for high temperatures and small densities.

First-order phase transitions in liquid crystals

In an ordinary simple isotropic liquid, both orientations and positions of molecules are random. In the liquid-crystal phase, positions of molecules may be random too, but their long axes are oriented on the average along a particular direction specified by the unit vector **n** that is called the director. The nematic phase of liquid crystals is characterized by broken rotation symmetry while the translational symmetry is not violated. In the case of a uniaxial liquid crystal (nematic or cholesteric) we may introduce the tensor

$$S_{ij} = S \left(n_i n_j - \frac{1}{3} \delta_{ij} \right) \tag{2.78}$$

for an order parameter that describes the ordering in the liquid crystal in terms of the scalar order parameter that is determined by the ratio of the number of oriented molecules to the total number of molecules in this condense matter, and the director **n** that determines the axis along which the most part of molecules are oriented. To specify the direction of the principal axis S_{ij} we write

$$S_{ij} = \frac{1}{2} (\langle 3cos^2\theta - 1 \rangle), \tag{2.79}$$

where θ is the angle between the molecule axis and the director **n**. The next step is to try to construct the Landau free energy for a nematic liquid crystal. The free energy density f should be invariant under all rotations of S_{ij} transformed as a tensor under the rotation group, f should be a function of only the scalar combination Tr S_{ij} that is invariant under rotations. To the fourth order in S_{ij}, the free energy may be presented as given by

$$f = \frac{1}{2}a\left(\frac{3}{2}\mathrm{Tr}\,S_{ij}^2\right) - c\left(\frac{9}{2}\mathrm{Tr}\,S_{ij}^3\right) + b\left(\frac{3}{2}\mathrm{Tr}\,S_{ij}^4\right). \tag{2.80}$$

The latter expression may be reduced to the condition

$$f = \frac{1}{2}aS^2 - cS^3 + bS^4. \tag{2.81}$$

In the general case, there should be two fourth-order terms proportional to $\mathrm{Tr}\,S_{ij}^3$ and $\mathrm{Tr}\,S_{ij}^4$. For a traceless tensor of a symmetric matrix, however, they are strictly proportional and should only contain the $(\mathrm{Tr}\,S_{ij}^2)^2$ term. Usually a behaves as $a = \alpha(T - T^*)$ and the other parameters are independent of temperature. The value of f at the minimum is greater than zero for high temperatures but vanishes for the critical temperature T_c that is higher than the temperature T^* for which the extremum at the origin develops negative curvature. Given that f is less than zero at the secondary minimum for all $T < T_c$, there occurs a phase transition with a discontinuous change in S for T_c, i.e., there is a first-order phase transition for T_c, and T^* is the limit of metastability of the isotropic phase since, for $T^* < T < T_c$, the origin is still a local minimum even though it is not a global minimum. The limit of metastability of the liquid crystal phase occurs for the temperature T^{**} when the secondary minimum disappears under heating. For the first-order phase transition temperature T_c and S_c the isotropic and liquid crystal phases can coexist. In terms of other variables, e.g., pressure or density, the two phases would coexist along a line rather than at a single point. The equation of state that determines T_s and S_c is given by

$$\frac{\partial f}{\partial S} = (a - 3cS + 4bS^2)S = 0. \tag{2.82}$$

The condition

$$f = \frac{1}{2}aS^2 - cS^3 + bS^4 = 0$$

give

$$S_c = \frac{c}{2b}, \quad a_c = \alpha(T - T^*) = \frac{c^2}{2b}. \tag{2.83}$$

The transition is of the first order hence it is associated with the lateral heat. The entropy per unit volume of the disordered phase vanishes in the mean-field theory. This result may be obtained from the free energy of the liquid crystal phase. In the lowest order in $a - a_c$, it is given by

$$f = \frac{1}{2}(a - a_c)S_c^2 = \frac{1}{2}(a - a_c)\left(\frac{c}{2b}\right)^2, \tag{2.84}$$

and then for the entropy density in the liquid crystal phase we have

$$s = -\frac{\partial f}{\partial T} = -\frac{1}{2}aS_c^2 = -\frac{1}{2}a\left(\frac{c}{2b}\right)^2. \tag{2.85}$$

Of course, there are other contributions to the entropy that are not included in the present model, nevertheless the total entropy is always positive. The latent heat absorbed under the transition from the liquid crystal to the isotropic phase may then be written as

$$q = -T_c s = \frac{1}{2}aT_c\left(\frac{c}{2b}\right)^2. \tag{2.86}$$

With an additional external field being present we have to take into account the interaction part of the free energy. In a magnetic field H this part takes the form

$$f_{\text{ext}} = -\frac{3}{2}\chi_a H^2 S, \tag{2.87}$$

where χ_a is the difference of the magnetic susceptibilities of the liquid-crystal molecules for the directions parallel and perpendicular to its long axis. Introducing the field conjugate to S. i.e., $h = (3/2)\chi_a H^2$, we obtain, as before, the susceptibility associated with S, i.e.,

$$\chi = \frac{\partial S}{\partial h} = (a - 6cS + 12bS^2)^{-1}. \tag{2.88}$$

The susceptibility χ diverges for $T = T^*$ as the temperature decreases in the isotropic phase. The first-order phase transition for $T_c > T^*$ cuts off this divergence. The temperature T^* is thus the limit of metastability of the isotropic phase. To find the limit of metastability of the liquid crystal phase T^{**}, we calculate the temperature for $\chi^{-1} = 0$ in the free-energy minimum when $S > 0$ satisfies the equation of state and thus obtain

$$a^{**} = \alpha(T^{**} - T^*) = \frac{9c^2}{16b} \qquad (2.89)$$

and $(T^{**} - T_c) = (c^2)/16ab$ strictly. The isotropic liquid crystal transition is a representative of phase transitions in which the order parameter contains a third-order invariant. One might expect in the general case that such transition should be of the first order. The Landau theory correctly predicts the qualitative properties of the first-order phase transitions. However, it cannot make detailed quantitative predictions because the order parameter is nonzero under the transition.

2.4 Dynamics of Metastable States

The dynamics of a system near the critical point is mainly governed by the fluctuations of the hydrodynamic modes, order parameter fields, and energy density. The slowness of relaxation thereof allows us to eliminate other degrees of freedom that get into local equilibrium for far shorter time. The equation for the order parameter relaxation is given by [65, 70, 73, 147]

$$\frac{1}{\gamma}\frac{\partial \varphi}{\partial t} = -\frac{\delta F}{\delta \varphi} + f, \qquad (2.90)$$

where γ is the kinetic coefficient; f is the external force that simulates the thermal ensemble. The variation of the field $\varphi(\mathbf{r}, t)$ at an arbitrary point for the time dt is equal to $\delta\varphi(\mathbf{r}, t) = (\partial\varphi)/(\partial t)dt$. Thus the total variation of the order parameter configuration energy is equal to

$$\delta F = \int \frac{\delta F}{\delta \varphi}\delta\varphi d\mathbf{r} = -\left\{\frac{1}{\gamma}\int \left(\frac{\partial\varphi}{\partial t}\right)^2 d\mathbf{r}\right\} dt. \qquad (2.91)$$

For a configuration associated with bubbles, the quantity $(\partial\varphi)/(\partial t)$ is appreciably different from zero only near the boundary. Within the context of the relation $\partial\varphi/\partial t = \dot{r}\nabla\varphi$ we find that

$$\left\{ \frac{1}{\gamma} \int \left(\frac{\partial\varphi}{\partial t} \right)^2 d\mathbf{r} \right\} dt = \dot{r}Sa^2 dt,$$

where $a^2 = [1/(\gamma S)] \int (\nabla\varphi)^2 d\mathbf{r}$ may be found for the equilibrium configuration of the field φ. On the other hand, we have

$$\delta F = \delta F_V + \delta F_S = \frac{2\xi\mu}{\lambda^{\frac{1}{2}}} \dot{r}Sdt + \frac{2\sigma}{r}\dot{r}Sdt, \tag{2.92}$$

where $(2\xi\mu)/\lambda^{1/2} = \varepsilon$ is the difference of field energies of the two phases and σ is the surface energy by definition. Making use of (2.91) and (2.92) we obtain an equation

$$\dot{r} = -\frac{2\sigma}{a^2}\left(\frac{1}{r} - \frac{1}{r_c} \right) \tag{2.93}$$

that describes the bubble size dynamics. Here r_c is its critical size as before, $r_c = (\mu^2/3\lambda^{1/2})\xi$.

Let us consider the distribution of sizes of new phase nuclei for an arbitrary time instant with allowance for the interaction of bubbles $(r \gg \lambda_0)$ with scale fluctuations $r \ll \lambda_0$ (λ_0 is the scale whose fluctuation amplitude is comparable to the spontaneous ordering). The concentration of such nuclei is small and hence the probability of their collisions may be neglected. Bubble interaction with the scale fluctuations $r \ll \lambda_0$ may be considered in terms of the effective random force. It was shown in [32, 44] that the equation for the bubble size dynamics is given by

$$\frac{dr}{dt} = -2\gamma\left\{ \frac{1}{r} - \frac{1}{r_c} \right\} + \chi. \tag{2.94}$$

In the latter equation an explicit expression is substituted for the coefficient a^2; the random force is equal to $\chi = (\partial\varphi/\partial r)^{-1}f$. For the case of Gaussian distribution of the force f with dispersion $2\theta\gamma$, the expression for the random force correlator χ may be written as

$$\langle \chi(r,t), \chi(r',t') \rangle = 2\theta\gamma \left| \int \left(\frac{\partial\varphi}{\partial r} \right)^2 d\mathbf{r} \right|^{-1} \delta(t - t') \qquad (2.95)$$

or, in terms of known quantities,

$$\langle \chi(r,t), \chi(r',t') \rangle = \frac{2\theta\gamma}{4\pi\sigma r^2} \delta(t - t'). \qquad (2.96)$$

The bubble size distribution for an arbitrary time instant is given by

$$W(r,t) = \int P(r,r_0,t)W(r_0)dr_0, \qquad (2.97)$$

where $P(r,r_0,t)$ is the probability that the bubble size is equal to r at time instant t for $r = r_0$ or $t = 0$. Distribution (2.97) satisfies the Fokker–Planck equation that represents the Langevin equation with the force correlator (2.95), i.e.,

$$\frac{\partial W}{\partial t} = \frac{\partial}{\partial r} \left\{ FW + D\frac{\partial W}{\partial t} \right\}, \qquad (2.98)$$

where

$$F(r) = 2\gamma \left\{ \frac{1}{r} - \frac{1}{r_c} \right\}, \quad D(r) = \frac{\theta\gamma}{4\pi\sigma r^2}.$$

The stationary solution of Eq. (2.98) is given by

$$W(r) = W_0 \exp \left\{ \frac{r^2}{D_0} \left[1 - \frac{2}{3} \frac{r}{r_c} \right] \right\}, \qquad (2.99)$$

where $D_0 = (\theta/4\pi\sigma)$. It is not difficult to see that this solution is in principle similar to the size distribution obtained by the statistical approach to the description of clusters in condensed media. The distribution grows infinite for $r > (3/2)r_c$; physically this means that the system transforms into a new phase state with infinite-size nucleus. For a time short as compared to the metastable state lifetime t_m, a stationary relaxation mode is established in the system — a small flow of nuclei towards the range of large sizes, i.e., $J = D(r)(\partial W/\partial t) + F(r)W = \text{const.}$

The mean time required for a nucleus of overcritical size to be formed is estimated to be equal to

$$t_m = J^{-1} = \frac{1}{W_0} \int_t^\infty \frac{1}{D(r)} \exp\left[\int_0^r \frac{F(r')}{D(r')} dr'\right] dr. \qquad (2.100)$$

To conclude, we consider the noise-induced first-order phase transitions. To do this, we have to assume that the governing parameter fluctuates. The fluctuations may be associated with the variations of the field external with respect to the order parameter that can simulate local changes of the microscopic interaction energy or temperature. In the general case, the noise-induced transitions are treated by the scheme with the relaxation described by the phenomenological equation of the form [76]

$$\dot{x} = f\alpha(x). \qquad (2.101)$$

The quantity x describes the state of the system. In our case, it corresponds to the order parameter with α being the governing parameter that depends on the state of the medium. In order to describe medium fluctuations, the parameter α contained in the phenomenological equation should be replaced by the random process $\alpha_t = \alpha + \Delta\xi_t$ with the external noise ξ_t whose average value and intensity Δ^2 vanish. In order to describe fast fluctuations of the medium, ξ_t may be treated to the white noise idealization. The main conclusion of the noise-induced transition theory [76] is that systems of the type (2.98) with the function $f_\alpha(x)$ linear with respect to the external parameter, $f_\alpha(x) = b(x) + \alpha g(x)$, may be simulated by a diffusion process caused by the external noise. In this case, evolution of the probability density $P(y, t \mid x, 0)$ is described by a Fokker–Planck-type equation, i.e.,

$$\frac{\partial}{\partial t} P(y, t) = -\frac{\partial}{\partial y}\left\{ f_\alpha(y) P(y, t) - \frac{\Delta^2}{2}\frac{\partial}{\partial y} g^2(y) P(y, t) \right\}. \qquad (2.102)$$

The stationary solution of the latter follows from

$$\frac{\partial}{\partial y} J = 0, \quad J = f P(y) - \frac{\Delta^2}{2}\frac{\partial}{\partial y} g^2(y) P(y).$$

It is given by

$$W(x) = \frac{W_0}{g^2(x)} \exp\left\{\frac{2}{\Delta^2} \int_0^x \frac{f(u)du}{g^2(u)}\right\} \qquad (2.103)$$

with the normalization condition

$$W_0^{-1} = \int \frac{dx}{g^2(x)} \exp\left\{\frac{2}{\Delta^2} \int_0^x \frac{f(u)du}{g^2(u)}\right\}.$$

We introduce an effective "probability" potential and rewrite the stationary probability density of the form

$$W(x) = W_0 \exp\left\{-\frac{2\nu}{\Delta^2}\right\}, \qquad (2.104)$$

where

$$\nu(x) = -\int_0^\infty \frac{f(u)du}{g^2(u)} + \frac{\Delta^2}{2} \ln g(u). \qquad (2.105)$$

Then the distribution extremum is determined by the solution of the equation

$$f(\tilde{x}) - \frac{\Delta^2}{2} g(\tilde{x}) g'(\tilde{x}) = 0 \qquad (2.106)$$

in which the prime denotes the derivative. This equation is basic for the analysis of the fast external noise effect on the stationary behavior of macroscopic nonlinear systems. It should be noted that a similar equation may be obtained by the general statistical approach to the description of thermodynamic behavior of the system taking into account fluctuations of the governing parameter. To do this, it is sufficient to add in (2.106) the free energy of interaction of the generalized force $g(x)$ with the governing parameter fluctuations that produce it, $F_{int} = (1/2)g(x)\Delta\xi$, and carry out averaging over the Gaussian distribution of these fluctuations. Then the partition function of the system is given by

$$Z \sim \int D\varphi \int D\xi \exp\left\{-\left[F(\varphi) + F_{int}(\varphi\xi) + \frac{1}{2}\xi^2\right]\right\}. \qquad (2.107)$$

Averaging over all probable fluctuation configurations reduces this expression to the standard form (2.106) with the new effective potential

$$V_{\text{eff}} = V(\varphi) - \frac{\Delta^2}{4} g^2(\varphi),\tag{2.108}$$

whose minimum is realized by the solution of the equation

$$\frac{dV(\varphi)}{d\varphi} - \frac{\Delta^2}{2} g(\varphi) \frac{dg(\varphi)}{d\varphi}\tag{2.109}$$

that is completely analogous to (2.106). The generalized force $g(\varphi)$ is by definition determined by the derivative of the term in the potential energy that depends on the governing parameter under consideration. For example, if the fluctuating parameter is μ^2, then $g(\varphi) = \varphi$; if λ fluctuates, then $g(\varphi) = \varphi^3$; having assumed that the fluctuating parameter is ξ we have $g(\varphi) = 1$. Thus, the general approach provides a possibility to analyze the influence of fluctuations of any governing parameter on the stationary macroscopic behavior of the system.

Chapter 3

Path Integration and Field Theory

3.1 Classical and Quantum Systems

In this chapter we consider the path-integration method that has recently proved to be rather efficient in statistical physics [7, 15]. Its efficiency owes to the fact that sometimes it is possible to work out a perturbation theory and obtain new qualitative results for a system lacking small parameters. The method is especially useful for many-particle systems within wide ranges of parameter variation. Moreover, the path-integration approach provides a possibility to obtain the values of thermodynamic quantities even in the vicinities of phase-transition points making no use of the complicated procedure of perturbation series summation.

We consider a physical system whose various configurations may be described in terms of continuous statistically independent variables $\varphi_l(\mathbf{r})$ that vary within the interval from $-\infty$ to $+\infty$. The subscript l takes arbitrary values and is simply a marker of the relevant field properties chosen for the description of the system. The energy of the system is a function of these variables, $E = E(\varphi_l)$, and the partition function may be formally given by the formula

$$Z = \prod_l \left\{ \int_{-\infty}^{\infty} \frac{d\varphi_l}{\sqrt{2\pi T}} \right\} \exp\left\{ -\frac{E(\varphi_l)}{T} \right\}, \qquad (3.1)$$

where $T = kT$ is the temperature times the Boltzmann constant. Thus, as usual, having determined the partition function we can start with the description of the thermodynamic properties of the system. The formal expression for the partition function is similar to the previous cases that required summation over all probable configurations, occupation numbers, etc. Now, however, the complete determined quantity is replaced by an arbitrary field that is intended to describe the system. The local properties of the system may be described in terms of correlation functions that are nothing but average values of some combinations of the fields introduced above, i.e.,

$$\chi^n(l_1, \ldots, l_n) = \langle \varphi_{l_1} \cdots \varphi_{l_n} \rangle$$
$$= Z^{-1} \prod_l \left\{ \int_{-\infty}^{\infty} \frac{d\varphi_l}{\sqrt{2\pi T}} \right\} \varphi_{l_1} \cdots \varphi_{l_n} \exp \left\{ -\frac{E(\varphi_l)}{T} \right\}.$$

$$(3.2)$$

In the general case, we have to take into account the interaction of the given field with the external fields. The simplest expression for the additional energy is $E_{\text{ext}} = -\sum_l j_l \varphi_l$, then the partition function $Z(j_l)$ depends on the field and determines the behavior of the system in this field. The arbitrary correlation function of the system may be written in the form

$$\chi^n(l_1, \ldots, l_n) = \langle \varphi_{l_1} \cdots \varphi_{l_n} \rangle = Z^{-1}(j_l) T^n \frac{\partial}{\partial j_{l_1}} \cdots \frac{\partial}{\partial j_{l_n}} Z(j_l). \quad (3.3)$$

The partition function $Z(j_l)$ is referred to as the generating function.

Now we come back to the calculation of the energy. In the general case, the energy of the system is given by the series

$$E(\varphi_l) = E^0 + \sum_l E_l^1 \varphi_l + \frac{1}{2!} \sum_{l,l'} \varphi_l V_{l,l'} \varphi_{l'}$$

$$+ \frac{1}{3!} \sum_{l,l'} V_{l,l',l''} \varphi_l \varphi_{l'} \varphi_{l''} + \cdots. \quad (3.4)$$

The latter energy representation contains both an independent part and probable many-particle interaction. Each addend is associated with the interaction of some type. For example, if φ_l are components of a vector, then E^1 is also a vector, and $V_{l,l'\ldots l^n}$ is a tensor of rank n. The energy in any representation should always be described by a scalar. It should be noted that expression (3.4) allegedly allows for the complete energy of the system taking into account any probable interactions. In what follows, however, we shall see that to do this we have to introduce many additional assumptions in the course of the calculation. In the general case, this problem has no solution. Nevertheless, the approach makes it possible to study the thermodynamic characteristics to much greater extent than with the previous methods. In the continuum case the energy of a many-particle system is given by

$$E(\varphi_l) = E^0 + \int E^1(\mathbf{r})\varphi(\mathbf{r})d\mathbf{r}$$

$$+ \frac{1}{2!} \int d\mathbf{r}d\mathbf{r}' \varphi(\mathbf{r})V(\mathbf{r}, \mathbf{r}')\varphi(\mathbf{r}') + \cdots . \tag{3.5}$$

In order to write the correlation functions, we have to add the energy of interaction with the external field, i.e., $E_{\text{ext}} = -\int d\mathbf{r}j(\mathbf{r})\varphi(\mathbf{r})$. The passage from the lattice to continuum is described by the relation $a^3 \sum_l j_l \varphi_l \sim \int d\mathbf{r}j(\mathbf{r})\varphi(\mathbf{r})$, where a is the lattice constant. Moreover, a scaling transformation of the field should be performed, i.e., $\varphi_l \equiv \sqrt{a^3}\varphi(\mathbf{r})$. The partition function in this representation is given by

$$Z = \prod_l \left\{ \int_{-\infty}^{\infty} \frac{d\varphi_l}{\sqrt{2\pi T/a^3}} \right\} \exp\left\{ -\frac{E(\varphi(\mathbf{r}))}{T} \right\}. \tag{3.6}$$

The partition function for the continuous case is given by the above expression with $a \to 0$. In this limiting case the quantity

$$\prod_l \left\{ \int_{-\infty}^{\infty} \frac{d\varphi_l}{\sqrt{2\pi T/a^3}} \right\} = \int D\varphi(\mathbf{r}) \tag{3.7}$$

is referred to as the path integral or continual integral. Thus we have a representation of the partition function in terms of the path integral over all probable field configurations in the system. The energy E^0 may be taken equal to zero since it just establishes the initial energy value, enters the partition function as an independent factor, and does not influence the state of the system.

First we consider the so-called free-field approximation that is formally described by the quadratic field-dependence of the energy. In the continuum case, it is given by the relation

$$
E(\varphi(\mathbf{r})) + E_{\text{ext}} = \frac{1}{2!} \int d\mathbf{r} d\mathbf{r}' \, \varphi(\mathbf{r}) V(\mathbf{r}, \mathbf{r}') \varphi(\mathbf{r}') - \int d\mathbf{r} j(\mathbf{r}) \varphi(\mathbf{r}).
$$

$$(3.8)$$

In the discrete case, we have

$$
E(\varphi_l) + E_{\text{ext}} = \frac{1}{2!} \sum_{l,l'} \varphi_l V_{l,l'} \varphi_l - \sum_l j_l \varphi_l.
$$

$$(3.9)$$

Then the partition function reduces to

$$
Z = \prod_l \left\{ \int_{-\infty}^{\infty} \frac{d\varphi_l}{2\pi T} \right\} \exp\left\{ -\frac{1}{T} \left(\frac{1}{2} \sum_{l,l'} \varphi_l V_{l,l'} \varphi_{l'} - \sum_l j_l \varphi_l \right) \right\}.
$$

$$(3.10)$$

If the energy is a quadratic function, then it may be written as a perfect square, i.e.,

$$
E(\varphi_l) + E_{\text{ext}} = \frac{1}{2} \sum_{l,l'} \left(\varphi_l - \sum_{l'} V_{l,l'}^{-1} \right) V_{l,l'} \left(\varphi_l - \sum_{l'} V_{l,l'}^{-1} \right)
$$
$$
- \frac{1}{2} \sum_{l,l'} j_l V_{l,l'}^{-1} j_{l'},
$$

$$(3.11)$$

where $V_{l,l'}^{-1}$ is the inverse matrix that satisfies the condition $V_{l,l'}^{-1} V_{l,l'} = \delta_{l,l'}$. Then we may introduce a new variable $\varphi_l' = \varphi_l - \sum_{l'} V_{l,l'}^{-1}$ and the partition function in terms of the new variables is given by

$$
Z = \prod_l \left\{ \int_{-\infty}^{\infty} \frac{d\varphi_l}{2\pi T} \right\} \exp\left\{ -\frac{1}{2T} \left(\frac{1}{2} \sum_{l,l'} \varphi_l' V_{l,l'} \varphi_{l'}' \right) \right\}
$$

$$
\times \exp\left\{ \frac{1}{2T} \sum_{l,l'} j_l V_{l,l'}^{-1} j_{l'} \right\}.
$$

$$(3.12)$$

To perform integration of this expression we need the inverse of the matrix $V_{l,l'}^{-1}$ that would have only diagonal elements. Thus, we perform an orthogonal transformation

$$V_{l,l'} = \sum_{l''} G_{l,l''} V_{l''} G_{l'',l'}^{-1} \tag{3.13}$$

along with the field transformation $\varphi_l'' = \sum_{l'} G_{l,l'}^{-1} \varphi_{l'}'$ that factorizes the product of exponentials $\prod_{l''} \exp\{-(1/2T) \sum_{l,l'} \varphi_l'' V_{l,l'} \varphi_{l'}''\}$ in the measure-preserving manner, i.e., $\prod_l d\varphi_l' = \prod_l d\varphi_l''$. The transformation Jacobian

$$\left| \frac{\partial \varphi_l''}{\partial \varphi_l'} \right| = \det G$$

is a unit operator since $G^{-1} = G^T$. In view of these observations, we reduce the partition function to the product of Gauss integrals

$$Z = \prod_l \left\{ \int_{-\infty}^{\infty} \frac{d\varphi_l''}{2\pi T} \right\} \exp \left\{ -\frac{1}{2T} \left(\frac{1}{2} \sum_{l,l'} \varphi_l'' V_l \varphi_{l'}'' \right) \right\}$$

$$\times \exp \left\{ \frac{1}{2T} \sum_{l,l'} j_l V_{l,l'}^{-1} j_{l'} \right\}. \tag{3.14}$$

Integrating each of these yields $1/\sqrt{V_l}$, and hence the partition function reduces to

$$Z = \left\{ \prod_l V_l \right\}^{-1/2} \exp \left\{ \frac{1}{2T} \sum_{l,l'} j_l V_{l,l'}^{-1} j_{l'} \right\}. \tag{3.15}$$

The product of eigenvalues V_l may be written as $\prod_l V_l = \det V$, i.e., it is reduced to a product of diagonal elements. Thus, the partition function is now given by

$$Z = (\det V)^{-1/2} \exp \left\{ \frac{1}{2T} \sum_{l,l'} j_l V_{l,l'}^{-1} j_{l'} \right\}. \tag{3.16}$$

Expression (3.15) is the basic formula of statistical physics in the path-integration method. We mention one more feature, though do not prove it. It could be shown that

$$(\det V)^{-1/2} = \exp\left\{-\frac{1}{2}\mathrm{Tr}\,\log V\right\}. \tag{3.17}$$

The latter equation makes it possible in some cases to write $\mathrm{Tr}\,\log V$ by a series of eigenvalues of the inverse interaction matrix, i.e.,

$$\mathrm{Tr}\,\log V = -\sum_n (-1)^n \mathrm{Tr}\,\frac{(V-1)^n}{n}.$$

An arbitrary correlation function in terms of the known expression for the partition function is given by

$$\chi^n(r_1,\ldots,r_n) = \langle \varphi_{l_1}\ldots\varphi_{l_n}\rangle = Z^{-1}T^n\frac{\partial}{\partial j_{r_1}}\cdots\frac{\partial}{\partial j_{r_n}}Z, \tag{3.18}$$

and for the average value of the field we have

$$\langle \varphi_l\rangle = \chi^1(l) = \sum_{l'} V_{l,l'}j_{l'}. \tag{3.19}$$

The two-point correlation function for particles located at different points of the space is then given by

$$\chi^2(r_1,r_2) = \langle \varphi_{l_1}\varphi_{l_2}\rangle = TV^{-1}(r_1,r_2) + \langle \varphi(r_1)\rangle\langle \varphi(r_2)\rangle. \tag{3.20}$$

We define the so-called contact correlation function in terms of the expression

$$\chi_c^2(r_1,r_2) = \chi^2(r_1,r_2) - \chi^1(r_1)\chi^1(r_2), \tag{3.21}$$

that physically implies that

$$\chi_c^2(r_1,r_2) = \langle(\varphi(r_1) - \langle \varphi(r_1)\rangle)\rangle \cdot \langle(\varphi(r_1) - \langle \varphi(r_1)\rangle)\rangle. \tag{3.22}$$

The values of this function are governed by the relation

$$\chi_c^2(r_1,r_2) = TV^{-1}(r_1,r_2), \tag{3.23}$$

that just determines the Green function of the integral equation

$$\int d\mathbf{r}'V(\mathbf{rr}')\varphi(\mathbf{r}') = j(\mathbf{r}). \tag{3.24}$$

We denote the two-particle correlation function by $\chi_0(r_1,r_2)$, then the partition function for the free field may be modified to the form that is completely equivalent to the previous one, i.e.,

$$Z = \exp\left\{-\frac{1}{2}\text{Tr}\log T\chi_0^{-1}\right\}\exp\left\{\frac{1}{2T^2}\int d^3r_1 d^3r_2 j(r_1)\right.$$

$$\left. \times \chi_0(r_1, r_2)j(r_2)\right\}. \tag{3.25}$$

This result formally solves the problem of statistical physics for the free field. We shall perform the specific calculation later on, and now we should like to draw the reader's attention to the fact that this representation of the partition function contains information on the structural parameters of the system. The formula may be employed to find the expressions for arbitrary thermodynamic functions of the system and, moreover, to obtain the correlation between any arbitrary points. For some types of interaction, however, the inverse operator for inverse-matrix eigenvalues is unknown. But once it is known, it is possible both to determine the partition function analytically and to derive the expressions for the thermodynamic quantities of interest.

Next, we give the most general definition of the partition function for quantum systems in terms of the functional integral. For particular issues, relevant representations are given with due consideration of each case. Formally, we may start from the expression for the classical partition function but take into account the specific nature of quantum variables. In the general case, the energy of the system φ_l depends both on the dynamical variables and on the canonical moments π_l that determine the Hamiltonian of the system, $H = H(\pi_l, \varphi_l)$. Once the Hamiltonian is known, the states of the system may be obtained from the Schrödinger equation

$$H\left(\frac{\hbar\partial}{i\partial\varphi_l}\right)\psi_m(\varphi_l) = E_m\psi_m(\varphi_l), \tag{3.26}$$

where $\psi_m(\varphi_l)$ is the wave function normalized in the sense of the scalar product

$$\left[\prod_l \int d\varphi_l\right]\psi_m^*(\varphi_l)\psi_{m'}(\varphi_l) = \delta_{m,m'}$$

and E_m are energy eigenvalues. For the field variables $\varphi(\mathbf{r})$ the above differential equation reduces to the functional one, i.e.,

$$H\left(\frac{\hbar}{i}\frac{\partial}{\partial\varphi(\mathbf{r})},\varphi(\mathbf{r})\right)\psi_m(\varphi(\mathbf{r})) = E_m\psi_m(\varphi(\mathbf{r})), \qquad (3.27)$$

where the wave function satisfies the orthogonality condition

$$\int D\varphi(\mathbf{r})\psi_m^*(\varphi(\mathbf{r}))\psi_{m'}(\varphi(\mathbf{r})) = \delta_{m,m'}. \qquad (3.28)$$

The partition function in statistical mechanics of quantum systems is given by

$$Z = \sum_m \exp\left\{-\frac{E_m}{T}\right\} \equiv \mathrm{Tr}\,\exp(-\beta\widehat{H}), \qquad (3.29)$$

hence we know the expressions for all thermodynamic functions of the system. Thus, we just have to obtain the partition function that may be written in terms of the integral over all probable paths [148]. To do this, we begin from some preliminaries. First of all, we find the density matrix of the quantum system. By definition, we have

$$\rho = \frac{\exp(-\beta H)}{\mathrm{Tr}\,\exp(-\beta H)}.$$

The density matrix as a function of the parameter β satisfies the equation

$$-\frac{\partial\rho}{\partial\beta} = H\rho.$$

To prove this relation we employ its expression in the energy representation, i.e., $\rho_{ij} = \delta_{ij}\exp(-\beta E_i)$. Differentiating this equation yields an equation

$$-\frac{\partial\rho_{ij}}{\partial\beta} = \delta_{ij}E_i\exp(-\beta E_i) = E_i\rho_{ij},$$

which just proves the above relation. The latter equation may be reduced to

$$\hbar(\partial\rho(u))/(\partial u) = H\rho(u) \qquad (3.30)$$

where the new variable $\beta\hbar \equiv u$ has time dimensions. The formal solution may be written as $\rho(u) = \exp(-Hu/\hbar)$. The "time" u may be divided into segments of length ε in a way that $(n\varepsilon = u)$ and then the density matrix reduces to a product, $\rho(u) = \rho_\varepsilon \rho_\varepsilon \ldots \rho_\varepsilon$. In the coordinate representation, the solution for the density matrix is given by

$$\rho(x, x', u) = \int \cdot \int \rho(x, x_{n-1}, \varepsilon) \ldots \rho(x_1, x', \varepsilon) dx_1 \ldots dx_{n-1}. \quad (3.31)$$

The physical meaning of this expression implies that a particle that moves from point x to x' passes a sequence of intermediate points $x_1, \ldots, x_{n-1}$ that determine its "path". The total amplitude of the probability $\rho(x, x', u)$ that a particle starting from the point x finishes its motion at the point x' is given by the sum over all probable "paths". When the time segment ε approaches zero, the number of integrations over intermediate variables grows infinite and the density matrix may be written in the symbolic form

$$\rho(x, x', u) = \int \cdot \int \Phi(x(u)) Dx(u), \quad (3.32)$$

where $Dx(u) = \lim_{n\to\infty} dx_1 \ldots dx_{n-1}$ and (as is shown in many books) we have $\Phi(x(u)) = \exp\{-(1/\hbar)A\}$. Here $A = \int_0^{\hbar\beta} L(u) du$ is action and L is the Lagrange function of the system. The integration range over "time" is limited because the boundary conditions are periodic and it is suggested that the whole behavior of the system may be considered in such a segment. Now, we may formally write the partition function for the field variables in terms of a path integral given by

$$Z = \int D\varphi(\mathbf{r}) \int \frac{D\pi(\mathbf{r})}{2\pi\hbar} \exp\left\{-\frac{1}{\hbar}A\right\}, \quad (3.33)$$

where the action for the discrete case is given by

$$A = \int_0^{\beta\hbar} du \sum_l \left[-i\pi_l \frac{\partial}{\partial u}\varphi_l + H(\varphi_l, \pi_l)\right], \quad (3.34)$$

and in the continuum case we have

$$A = \int_0^{\beta\hbar} du \int d\mathbf{r} \left[-i\pi(\mathbf{r}) \frac{\partial}{\partial u} \varphi(\mathbf{r}) + H(\varphi(\mathbf{r}), \pi(\mathbf{r})) \right]. \qquad (3.35)$$

The formulas thus obtained are referred to as Euclidean expressions because they are defined in the ordinary space.

3.2 Saddle-Point Method or Stationary-Phase Method

Calculating path integrals is a rather complicated task, especially in the cases when the energy cannot be written in quadratic form. Such integrals can be calculated exactly only in some special cases. Thus, we need an approximate calculation method. Among the approaches known in quantum mechanics, the stationary-phase method is rather efficient. Its name will be explained somewhat later. The method is used in quantum mechanics to calculate the matrix of transitions between different states making allowance for all probable trajectories of a quantum particle. It is very important since it is inherently associated with the quasi-classical approximation. The general relevant problem in quantum mechanics is to calculate the integrals of the type

$$I(\lambda) = \int_a^b dx f(x) \exp\left[i\lambda S(x)\right] \qquad (3.36)$$

for the cases when the parameter entering the exponent takes large values. Actually $S(x)$ is the action, but in some cases it is referred to as the phase function; $f(x)$ is a function that is continuous in some definite interval and has arbitrarily many derivatives. The stationary point is determined by the extremum condition $\partial S(x)/\partial x = 0$. When the parameter tends to infinity, $\lambda \to \infty$, then, due to fast oscillations of the exponential, the greatest contribution in the integral is the small endpoints and the near-end points as well as by the stationary points of the phase function, provided they belong to the above-mentioned definite interval. Here we give the result that makes it possible to find the asymptotics of the integral (3.36) for $\lambda \to \infty$. We suppose that the point x_c that should be found from the extremum condition $\partial S(x)/\partial x = 0$ is the only stationary point in the integration range. If there are several such points, each of these contributes to the integral. Moreover, these points should not be degenerate, i.e., the second derivative of the action at stationary points should not be equal to zero. Under these conditions, the leading term of the integral asymptotics is given by

$$I(\lambda) \sim \left(\frac{2\pi}{\lambda |S''(x_c)|} \right)^{1/2} f(x_c) \exp \left\{ i\lambda S(x_c) + \frac{i\pi}{4} \text{sign} S''(x_c) \right\},$$

$$(3.37)$$

where $S''(x_c)$ is the second derivative of the action for the coordinate at the stationary point. To verify the last formula we expand the action, also referred to as the phase function, in Taylor series in the vicinity of the stationary point, i.e.,

$$S(x) \simeq S(x_c) + S'(x_c)(x - x_c) + S''(x_c)(x - x_c)^2 + \cdots, \quad (3.38)$$

and retain only the quadratic term. This expression for the path integral may be easily extended to the many-dimensional case. In case we need a more general form of the integral, i.e.,

$$I(\lambda) = \int_a^b d\mathbf{x} f(x) \exp(i\lambda S(\mathbf{x})), \quad (3.39)$$

where $\mathbf{x} = (x_1 \ldots x_n)$ is the set of all coordinates, then we have $d\mathbf{x} = dx_1 \ldots dx_n$, where n is the space dimension. The phase function may be expanded in the vicinity of the stationary point similar to the one-dimensional case, then we have

$$S(\mathbf{x}) \simeq S(x_c) + \sum_i S_i'(\mathbf{x_c})(\mathbf{x_i} - \mathbf{x_c})$$

$$+ \sum_{i,j} S_{i,j}''(\mathbf{x_c})(\mathbf{x_i} - \mathbf{x_c})(\mathbf{x_j} - \mathbf{x_c}) + \cdots. \quad (3.40)$$

Now the matrix

$$S''_{i,j}(\mathbf{x_c}) = \left(\frac{\partial^2 S(\mathbf{x})}{\partial x_i \partial x_j}\right)_{\mathbf{x}=\mathbf{x_c}}$$

is real and symmetric. Hence the leading term of the path integral may be written in the form

$$I(\lambda) = \left(\frac{2\pi}{\lambda}\right)^{n/2} f(\mathbf{x_c})|\mathrm{Tr}\, S''_{i,i}|^{-1/2} \exp\left\{i\lambda S(\mathbf{x_c}) + \frac{i\pi}{4}\mathrm{sign}S''_{i,i}\right\}$$

(3.41)

(the notation sign $S''_{i,i}$ implies the number of positive and negative eigenvalues of the relevant matrix).

Now, we have to determine the asymptotics of the path integral in quantum mechanics. The integral $I(\lambda)$ is an arbitrary Feynman path integral of a quantum particle with the parameter λ of the form $1/\hbar$. We begin with the simplest problem of one-dimensional particle motion in the potential field $V(x)$. Now we expand the action functional $S(x,t)$, that is always contained in the definition of the path integral, close to the classical path $x_c(t)$. The particle coordinate may be written as $x(t) = x_c(t) + \delta x$, where δx describes the deviation of the particle trajectory from the classical path. Note that it vanishes at the ends of the given segment. Then the particle action functional reduces to

$$S(x,t) = \int_{t_0}^{t} dt \left\{\frac{m(\dot{x})^2}{2} - V(x)\right\} \int_{t_0}^{t} dt \left\{\frac{m(\dot{\delta x})^2}{2} - \frac{1}{2}\frac{\partial^2 V(x)}{\partial x^2}\delta x^2\right\}$$

$$+ \int_{t_0}^{t} dt \left\{m\dot{x}\delta x - \frac{\partial V(x)}{\partial x}\delta x\right\}.$$

(3.42)

Integration by parts provides evidence that the last addend in the right-hand part of the action functional is identically equal to zero because $S'(x,t) = 0$ in the classical path $x_c(t)$. The final expression for the phase functional is given by

$$S(x,t) = S_{\text{cl}}(x_c(t)) + \frac{m}{2} \int_{t_0}^{t} dt \delta x \widehat{U(t)} \delta x \qquad (3.43)$$

with the operator

$$\widehat{U(t)} = -\frac{\partial^2}{\partial t^2} - \frac{1}{m} \left(\frac{\partial^2 V}{\partial x^2} \right)_{x(t)=x_c(t)}. \qquad (3.44)$$

Thus we see that an arbitrary path integral over all probable paths may be described in terms of the asymptotics

$$\int Dx(t) \exp\left[\frac{i}{\hbar} S(x(t))\right] \cong A |\text{Tr}\widehat{U(t)}|^{-1/2} \exp\left(\frac{i}{\hbar} S(x_c(t))\right),$$

$$(3.45)$$

where A is a constant that contains all the factors that arise in the course of integration.

Now we return to the derivation of the partition function in terms of the path integral over all probable configurations of the fields associated with the system. First, we consider the one-dimensional case. We have already mentioned that the partition function of the system may be represented by the standard expression

$$Z = \int \frac{dx}{\sqrt{2\pi T}} \exp\left(-\frac{f(x)}{T}\right). \qquad (3.46)$$

The latter expression reproduces the quantum-mechanics path integral over all probable trajectories. The most important distinction is that all functions entering the representation of the partition function are real and hence we may calculate the partition function by the method similar to the previous case. The parameter that can take large values is the reciprocal temperature, hence the partition function asymptotics is associated with the low-temperature states of the system. Now we expand the function $f(x)$ in the vicinity of the maximum or minimum point of this function that is determined by the condition $\left(\frac{\partial f(x)}{\partial x}\right)_{x=x_c} = 0$ in series of deviation from this point, i.e.,

$$f(x) = f(x_c) + \frac{1}{2}f''(x_c)(x - x_c)^2 + \cdots,$$

and substitute it in the path integral. Thus, we have

$$Z = \frac{1}{\sqrt{f''(x)}} \exp\left\{-\frac{f(x_c)}{T}\right\}. \tag{3.47}$$

This result may be extended to the path integral in the space of arbitrary dimension. The partition function in terms of the field variables in a D-dimensional space is given by

$$Z = \int \frac{d\varphi(\mathbf{x})}{\sqrt{2\pi T}} \exp\left\{-\frac{E(\varphi(\mathbf{x}))}{T}\right\}. \tag{3.48}$$

For temperatures $T \to 0$, the dominant contribution is given by the configurations $E\varphi_c(\mathbf{x})$ corresponding to the extremum

$$\frac{\partial E(\varphi(\mathbf{x}))}{\partial \varphi(\mathbf{x})} = 0.$$

For the three-dimensional space, we introduce the notation

$$\frac{\delta^2 E(\varphi(\mathbf{x}))}{\delta\varphi(\mathbf{r_1})\delta\varphi(\mathbf{r_2})} = D(\mathbf{r_1 r_2}),$$

and then the partition function in this approximation is given by

$$Z = \exp\left\{-\frac{E(\varphi_c)}{T}\right\}(\det D)^{1/2}$$

$$\times \exp\left\{\frac{1}{2}\left[\int d\mathbf{r_1}d\mathbf{r_2}j(\mathbf{r_1})D^{-1}(\mathbf{r_1},\mathbf{r_2})j(\mathbf{r_2})\right]\right\}. \tag{3.49}$$

Thus, we have obtained an expression for the partition function in the low-temperature limiting case. This method is referred to as the *saddle-point method*. The name owes to the observation that the dominant contribution is associated with the configurations determined by the minimum-energy condition whereas all minimum values lie in the lowest points of the field-dependence relief of the energy. The same observation concerns quantum mechanics. Given that the first derivative of action vanishes at some points, the phase of the wave function is not changed, and thus we have the name.

In this section we give the simplest results that can be obtained by the path-integration method. Actually, it is a highly efficient approach that in many cases makes it possible to derive the final expressions for the partition function that could not be obtained by other methods. The greatest disadvantage is the lack of knowledge of inverse operators for many interactions. Exact results can be obtained only for the cases when the path integral can be reduced to a Gaussian-type integral, i.e., only for the quadratic dependence on the field functions.

3.3 Construction of the Field Theory

The statistical and thermodynamic description of a system of interacting particles has been the basis for understanding dense matter for many years. Within the framework of this approach the phase transitions in a system of particle interaction from a gas phase to liquid, and also from liquid to solid are described [123–132]. In Chapter 1 we studied the description of simple models of statistical physics of interacting particles and used the mean-field theory based on the phenomenological expansion of the free energy in the power series of the order parameter. In order to include fluctuations that become important for critical temperatures, it is necessary to have a microscopic description of the partition function. We have to start from the microscopic Hamiltonian expressed either in terms of quantum-mechanics operator or in terms of classical dynamical variables and to make an attempt to evaluate the partition function density. This is a very complicated procedure. It is difficult to carry out in the vicinity of the second-order phase transition, when the correlation length diverges. To study the critical properties of the phase transitions, we have to introduce a more appropriate semi-phenomenological field theory, where the trace operation is an integral over all probable values at all points of space of the local order parameter considered as the continuous classical field. The field theory should describe the system far from any phase transition.

The partition function involves a trace over all probable states of the system. It should contain information about the order parameter. The order parameter field $\varphi(\mathbf{x})$, like the Hamiltonian, is a quantum-mechanics operator or a function of the classical dynamical variables. The first step is to divide the system into many cells with dimensions large compared to any microscopic length such as the interparticle spacing or the range of the interparticle potential. In the center of each cell we introduce the order parameter field. Each cell contains many particles. The field $\varphi(\mathbf{x})$ may be treated as a continuous classical variable that can vary from cell to cell. The states of the system may be now specified by this field and have the effective energy $H(\varphi)$. The partition function in this case is integral (functional integral) over all probable configurations of the field $\varphi(\mathbf{x})$ at all positions $\mathbf{x}$, i.e.,

$$Z = \int D\varphi(\mathbf{x}) \exp\left\{-\beta[\tilde{H} - \int d^d x h(\mathbf{x})\varphi(\mathbf{x})]\right\}, \qquad (3.50)$$

where $1/kT$ is the reciprocal temperature and $h(\mathbf{x})$ is the external field. The integration over $\varphi(\mathbf{x})$ is often called a path integral because in the one-dimensional case it is an integral over all probable spatial paths of the field. In the case of a classical system, the partition function may be obtained by introducing the delta function

$$Z = \int D\varphi(\mathbf{x}) \prod_x \delta(\varphi(\mathbf{x}) - \overline{\varphi(\mathbf{x})}) \exp\left\{-\beta[H - \int d^d x h(\mathbf{x})\varphi(\mathbf{x})]\right\},$$

$$\qquad (3.51)$$

where $\overline{\varphi(x)}$ is the mean-field value and

$$\exp\left\{-\frac{\tilde{H}}{kT}\right\} = \text{Tr} \prod_x \delta(\varphi(\mathbf{x}) - \overline{\varphi(\mathbf{x})}) \exp\left\{-\frac{H}{kT}\right\}. \qquad (3.52)$$

The effective energy $\tilde{H}$ is usually called the Hamiltonian in statistical mechanics. It is closed to the action in the quantum field theory. Usually the original microscopic Hamiltonian is not considered in detail, but the effective energy determines the weight assigned to the configuration in the functional integral representation of the partition function. The phenomenological form of H may be taken similar to the Landau form for the mean-field theory. There should be present a local part depending only on the order parameter and a part that describes the spatially uniform order parameter, i.e.,

$$H = \int d^d x f(\varphi(\mathbf{x})) + \frac{1}{2} \int d^d x c (\nabla \varphi(\mathbf{x}))^2. \tag{3.53}$$

This Hamiltonian provides a description of the energy associated with long-wavelength slow spatial variations of $\varphi(\mathbf{x})$. It does not provide a realistic description of the short-wavelength distortion. One approach to the solution of this problem is to introduce the wave number greater than the cutoff $\Lambda \sim (2\pi/a)$ where a is the length of the order of magnitude of the range of interparticle interaction. The second method is to take into account the short-wavelength fluctuations. It may be employed when we consider an additional term proportional to $(\nabla^2 \varphi(\mathbf{x}))^2$. The definition of the functional integral in the partition trace requires considering the continuum limit of the theory with the field φ_l defined on N sites l of a d-dimensional lattice. The partition function for the lattice model may be written in the form

$$Z = \prod_l d\varphi_l \exp(-\beta H_l), \tag{3.54}$$

where

$$H_l = \sum_l f_l(\varphi_l) + \frac{1}{2} G_{l,m}(\varphi_l - \varphi_m) \tag{3.55}$$

with f_l being the power series expansion and $G_{l,m}$ having finite range equal to the lattice constant. The continuum limiting case of the lattice model may be obtained when the lattice constant tends to zero

$$v_o \sum_l \rightarrow \int d^d x, \quad \sum_l f_l(\varphi_l) \rightarrow \int d^d x \tilde{f}(\varphi(\mathbf{x})),$$

$$\frac{1}{2} G_{l,m}(\varphi_l - \varphi_m) \rightarrow \frac{1}{2} \int d^d x c (\nabla \varphi(\mathbf{x})),$$

where

$$V = N v_0, \quad \widetilde{f} = v_0^{-1} f_l, \quad c = \frac{1}{dv_0} \sum_l \mathbf{R}_l^2 G_{l,0}.$$

The functional integral of the continuum partition function is given by the formal expression

$$\int D\varphi(\mathbf{x}) = \lim_{v_0 \to 0} \prod_l \int d\varphi_l.$$

Finally, the lattice model becomes continuum. Both lattice and continuum field theories may be applied extensively in condensed matter physics.

3.4 Hubbard–Stratonovich Transformation

The phenomenological continuum gradient field theory requires introducing the functional integral representation for the partition function of strongly constrained models such as Ising model, harmonic model, and n-vector model. In all these models, it is possible to study the properties near the second-order phase transition. In this section, we consider a method for obtaining the general representation of the partition function as a continuum integral over all probable configurations that can be realized in a system of interacting particles that may be described in terms of a quadratic Hamiltonian. For the sake of simplicity we consider only the Ising model, though generalization is possible to any microscopic Hamiltonian that may be expressed in a quadratic form. The starting point is the Ising Hamiltonian given by the sum over all sites, i.e.,

$$H = -\frac{1}{2} \sum_{l,m} J_{lm} S_l S_m, \tag{3.56}$$

where $S_l = \pm 1$, J_{lm} is the exchange integral that in the simplest model is nonzero unless l and m are nearest neighbor sites. In this case, we may write

$$\exp\left\{\frac{1}{2}\sum_{l,m} K_{lm}S_l S_m\right\}$$

$$= A\prod_l \int d\varphi_l \exp\left\{-\frac{1}{2}\sum_{l,m}\varphi_l K_{lm}^{-1}\varphi_m + \sum_l \varphi_l S_l\right\}, \quad (3.57)$$

where $K_{lm} \equiv \beta J_{lm}$ and $A = (2\pi)^{-N/2}(\det K)^{-1/2}$. The Ising-model partition function in this presentation is given by

$$Z = A\prod_l \int d\varphi_l \exp\left\{-\frac{1}{2}\sum_{l,m}\varphi_l K_{lm}^{-1}\varphi_m\right\} \operatorname{Tr}\left\{\sum_l (\varphi_l + \beta h_l)S_l\right\}.$$

$$(3.58)$$

The trace in this expression is over the spin variables. If the spin variable appears in a linear manner rather than quadratic, then it is possible to obtain the exact evaluation of the trace over S_l, this feature being the great advantage of the Hubbard–Stratonovich transformation. The integral over φ_l is nontrivial. It is easy to show that

$$\langle S_l\rangle = \frac{\partial \ln Z}{\partial \beta h_l} \quad \text{and} \quad \langle S_l\rangle = \sum_l K_{lm}\langle \varphi_l\rangle$$

by integration after shifting the derivation with respect to βh_l to the derivation with respect to φ_l. The unconstrained field φ_l is proportional to the constrained field S_l. The dependence of the partition function on h_l may be transformed to a more suitable form by changing the variables. For $\psi_l = \varphi_l + \beta h_l$ we obtain the partition function Z given by the expression

$$\prod_l \int d\varphi_l \exp\left\{-\frac{1}{2}\sum_{l,m}(\psi_l - h_l)K_{lm}^{-1}\psi_m\right.$$

$$\left.-\frac{1}{2}\sum_{l,m}\beta h_l K_{lm}^{-1}\beta h_m - \sum_l \beta f_l \psi_l\right\}, \quad (3.59)$$

where

$$\beta f_l(\psi) = -\ln \mathrm{Tr}\,(\exp S\psi) = -\ln(2\cosh\psi). \qquad (3.60)$$

This is the exact solution of the lattice model and in the continuum limiting case the procedure may be carried out in a manner similar to the previous one. The additional field introduced in the course of calculation of the partition function contains information about many configurations that have been described in terms of the spin.

3.5   The Mean-Field Theory and the Functional Integral

So, we described the field theory for the case when both the thermodynamic potential and its configuration potential may be calculated. To calculate the partition function Z we have to take into account all probable configurations in the functional space. The general way to consider the configurations with the greatest contributions in Z is to minimize $\beta(H - \int d^d x h(\mathbf{x})\varphi(\mathbf{x}))$. In the saddle point of the functional integral, the dominant term is the one that contains $\varphi(\mathbf{x}) = \overline{\varphi}(\mathbf{x})$ that is determined by the equation

$$\frac{\delta H}{\delta \varphi(\mathbf{x})}\bigg|_{\varphi(\mathbf{x})=\overline{\varphi}(\mathbf{x})} = h(\mathbf{x}). \qquad (3.61)$$

The mean-field theory consists of approximating Z by its contribution from the saddle point path only. In the mean-field approximation, the partition function may be written in the form:

$$Z = \exp\left\{-\beta H(\overline{\varphi(\mathbf{x})}) - \int d^d x h(\mathbf{x})\overline{\varphi(\mathbf{x})}\right\}. \qquad (3.62)$$

Corrections to the mean-field theory may be obtained by expanding H in powers of $\delta\varphi(\mathbf{x}) = \varphi(\mathbf{x}) - \overline{\varphi}(\mathbf{x})$. In the general case, due to fluctuations $\langle\varphi(\mathbf{x})\rangle$ should differ from $\overline{\varphi}(\mathbf{x})$ in the ordered phase or in the presence of an external field. Expanding the powers of the exponent to the second order in $\delta\varphi(\mathbf{x})$ yields

$$H - \int d^d x h(\mathbf{x})\varphi(\mathbf{x}) = H\big(\langle\varphi(\mathbf{x})\rangle\big) - \int d^d x h(\mathbf{x})\langle\varphi(\mathbf{x})\rangle + H', \quad (3.63)$$

where

$$\beta H' = \frac{1}{2}\int d^d x d^d x' \delta\varphi(\mathbf{x}) G_0^{-1}\delta\varphi(\mathbf{x}) \qquad (3.64)$$

is the harmonic correction to the saddle-point Hamiltonian and

$$G_0^{-1}(\mathbf{x},\mathbf{x}') = \frac{\delta\beta H}{\delta\varphi(\mathbf{x})\delta\varphi(\mathbf{x}')} = \beta\Big\{a + 12b\big(\langle\varphi(\mathbf{x})\rangle\big)^2 - k\nabla^2\Big\}\delta(\mathbf{x}-\mathbf{x}')$$

$$(3.65)$$

is the inverse of the mean-field correlation function where the term φ^4 is finite. Harmonic fluctuations about the mean-field or saddle-point solution are governed by the mean-field order parameter. The lowest-order fluctuation correlation to the mean-field free energy may be calculated by applying the Gaussian integration over $\delta\varphi(\mathbf{x})$. Thus, we obtain

$$F - F_{mf} = \frac{1}{2}T\mathrm{Tr}\left\{\frac{G_o^{-1}v_0}{2\pi}\right\}. \qquad (3.66)$$

The second-order term in this expansion gives the one-loop correction to the inverse correlation function that in the disordered phase with $\langle\varphi(\mathbf{x})\rangle = 0$ is given by

$$G^{-1}(\mathbf{x},\mathbf{x}') = G_0^{-1} + \frac{1}{3}\mathrm{Tr}\, G_0 \frac{\delta G_0^{-1}}{\delta\varphi(\mathbf{x})\delta\varphi(\mathbf{x}')}. \qquad (3.67)$$

For φ^4 theory, this relation reduces to

$$G^{-1}(\mathbf{x},\mathbf{x}') = G_0^{-1}(\mathbf{x},\mathbf{x}') + 12b\delta(\mathbf{x}-\mathbf{x}')G_0(\mathbf{x},\mathbf{x}') \qquad (3.68)$$

that in the disordered phase yields $\langle\varphi(\mathbf{x})\rangle = 0$, and

$$TG^{-1}(\mathbf{q},a) = a + kq^2 + 12b\int \frac{d^d q}{(2\pi)^d}\frac{kT}{a + kq^2}. \qquad (3.69)$$

One-loop corrections to the fourth and higher-order terms of the expansion of the free energy in power series of $\langle\varphi(\mathbf{x})\rangle$ may be calculated in a similar way. In what follows we show how fluctuations lead to the breakdown of the mean-field theory. We see that replacing the contribution of the partition function from the most probable path with a spatially uniform order parameter is equivalent to the mean-field theory. When $h(\mathbf{x})$ is spatially uniform, then the saddle-point path yielding the mean-field theory is the one with a spatially uniform $\langle\varphi(\mathbf{x})\rangle$. This approximation is very good as long as the contribution of the spatially nonuniform paths to the partition function is unimportant, in such a case when the energy

$$H_\nu = \frac{1}{2}\int d^d x k(\nabla\varphi)^2$$

is associated with the nonuniform paths. It is possible to estimate the magnitude of this energy provided we bear in mind that $(\langle\varphi(\mathbf{x})\rangle)^2 = a/4b$ at the saddle point and the spatial correlation only extends up to the distance of the order of magnitude of

$$\xi = \left(\frac{k}{|a|}\right)^{-1/2}.$$

After this, we estimate that

$$H_\nu = \frac{1}{2}\int d^d x k(\nabla\varphi)^2 \sim \frac{1}{2}k\xi^{d-2}(\langle\varphi(\mathbf{x})\rangle)^2 = \frac{k^2}{8b}\xi^{d-4}. \qquad (3.70)$$

In the mean-field theory this energy should be greater than T and $(k^2/8b)\xi^{d-4} > T$. The Ginzburg temperature also indicates when the perturbation theory will breakdown. We consider the one-loop expression for

$$\frac{dT G^{-1}(\mathbf{q}a)}{da} = 1 - 12bT\int \frac{d^d q}{(2\pi)^d}\frac{1}{a + kq^2}. \qquad (3.71)$$

The perturbation term in this expression is proportional to b as $(bT/k^2)\xi^{4-d}$ and becomes comparable to one for Ginzburg temperature T_G.

In this section, we consider the self-consistent field approximation. In this case φ^4 in the harmonic free energy is approximately given by $(1/2)a\varphi 2 + 6\langle\varphi^2\rangle\varphi^2$. The correlation function $G(\mathbf{q})$ may be calculated using the theorem on the uniform distribution of energy per one degree of freedom. We have

$$kTG^{-1}(\mathbf{q}) = a + q^2 + 12b\langle\varphi^2\rangle \qquad (3.72)$$

and thus we obtain the self-consistent equation in the form

$$\langle\varphi^2\rangle \equiv G(\mathbf{x}, \mathbf{x}) = \int \frac{d^d q}{(2\pi)^d} G(\mathbf{q}) = \int \frac{d^d q}{(2\pi)^d} \frac{kT}{a + kq^2 + 12b\langle\varphi^2\rangle}.$$
$$(3.73)$$

This equation provides a very good description of the behavior of the order parameter in the long-wavelength approximation. For large q, higher powers in the expansion of this wavelength are required, or we have to introduce some cutoff Λ of the order of the correlation length.

Chapter 4

Peculiarity of Calculation
of Some Models

4.1 Special Cases of the Calculation of Path Integrals

First of all, we consider a system of N noninteracting particles. Its Hamiltonian is equal to $H = \sum_i (p_i^2/2m)$ and the partition function is given by

$$Z = \int D\mathbf{r_i} \int \frac{d\mathbf{r_i}}{\sqrt{(2\pi\hbar^3)/m}} \exp\left\{-\frac{m}{2\hbar}\sum_{i=1}^{N}\int_0^{\beta\hbar}\left(\frac{d\mathbf{r_i}}{du}\right)du\right\}, \quad (4.1)$$

$$Z = \int D\mathbf{r_i} \left(\frac{m}{2\pi\hbar^2\beta}\right)^{3N/2} \prod_i^N \exp\left\{-\frac{m}{2\hbar^2\beta}(\mathbf{r_i} - \mathbf{r_i'})\right\}$$

$$= \left(\frac{m}{2\pi\hbar^2\beta}\right)^{3N/2} V^N. \quad (4.2)$$

This expression may be derived in another way. The above equation for the density matrix in the three-dimensional space may be written as

$$-\frac{\partial\rho}{\partial u} = -\frac{\hbar^2}{2m}\Delta\rho, \quad (4.3)$$

where Δ is the three-dimensional Laplace operator. The equation thus obtained is the diffusion equation in terms of complex variables. Its solution may be written in the form

$$\rho = \left(\frac{m}{2\pi\hbar^2\beta}\right)^{3/2} \exp\left\{-\frac{m}{2\hbar^2\beta}(\mathbf{r} - \mathbf{r'})^2\right\}, \quad (4.4)$$

where the factor is chosen in a way to satisfy the relation $\rho(\mathbf{r}, \mathbf{r}') = \delta(\mathbf{r} - \mathbf{r}')$. Then the partition function reduces to

$$Z = \left(\frac{m}{2\pi\hbar^2\beta}\right)^{3N/2} \int \prod_i D \exp\left\{-\frac{m}{2\hbar^2\beta}(\mathbf{r}_i - \mathbf{r}_i')^2\right\}$$

$$\mathbf{r}_i = \left(\frac{m}{2\pi\hbar^2\beta}\right)^{3N/2} V^N.$$

(4.5)

A similar but more demonstrative result for free particles may be obtained provided that we begin the description from one free particle with one degree of freedom. In this case $H = p/2m$ and the equation for the density matrix reduces to

$$-\frac{\partial\rho}{\partial u} = -\frac{\hbar^2}{2m}\frac{\partial^2}{\partial x^2}\rho.$$

(4.6)

The solution of this equation is similar to the solution of the one-dimensional diffusion-type equation, i.e.,

$$\rho(x, x', \beta) = \left(\frac{m}{2\pi\hbar^2\beta}\right)^{1/2} \exp\left\{-\frac{m}{2\hbar^2\beta}(x - x')^2\right\}.$$

(4.7)

Then the partition function for a one-dimensional system of length L is equal to

$$Z = \int \rho(x, x)dx = L\left(\frac{m}{2\pi\hbar^2\beta}\right)^{1/2}.$$

(4.8)

Hence it is obvious that in the three-dimensional case the partition function of a system of N particles is given by

$$Z = V^N \left(\frac{m}{2\pi\hbar^2\beta}\right)^{3N/2}.$$

(4.9)

Now we consider a linear harmonic oscillator. The Hamiltonian of a linear harmonic oscillator with frequency ω is given by

$$H = \frac{p^2}{2m} + \frac{m\omega^2 x^2}{2}.$$

(4.10)

Thus the equation for ρ reduces to

$$-\frac{\partial \rho}{\partial \beta} = -\frac{\hbar^2}{2m}\frac{\partial^2}{\partial x^2}\rho + \frac{m\omega^2}{2}x^2\rho. \qquad (4.11)$$

We introduce the new variables $\xi = \sqrt{(m\omega/\hbar)}x$ and $\tau = (\hbar\omega/2)\beta$, then the equation is simplified and we have

$$-\frac{\partial \rho}{\partial \tau} = -\frac{\partial^2 \rho}{\partial \xi^2} + \xi^2\rho. \qquad (4.12)$$

The latter equation with the initial condition $\rho = \delta(x - x')$ for $\tau = 0$ is solved in [148]. The solution is given by

$$\rho(x, x', \beta) = \sqrt{\frac{m\omega}{2\pi\hbar\sinh 2\tau}}\exp\left\{-\frac{m\omega}{2\hbar\sinh 2\tau}\right.$$
$$\left. \times \left[(x^2 + x'^2)\cosh 2\tau - 2xx'\right]\right\}. \qquad (4.13)$$

In the particular case $x = x'$, we have

$$\rho(x, x, \beta) = \sqrt{\frac{m\omega}{2\pi\hbar\sinh 2\tau}}\exp\left\{-\frac{m\omega}{2\hbar\sinh 2\tau}x^2\tanh 2\tau\right\}. \qquad (4.14)$$

This expression yields the partition function

$$Z = \int \rho(x, x)dx = \sqrt{\frac{m\omega}{2\pi\hbar\sinh 2\tau}}\sqrt{\frac{\pi\hbar}{m\omega\tanh 2\tau}} = \frac{2}{\sinh 2\tau}, \qquad (4.15)$$

and hence the free energy is given by

$$F = \ln Z = \ln\frac{2}{\sinh 2\tau} = \frac{\hbar\omega}{2} + kT\ln\left[1 - \exp\left(-\frac{\hbar\omega}{kT}\right)\right]. \qquad (4.16)$$

Thus, we have derived the formulas that had been known earlier by using the equation for the density matrix.

In the general case of field representation of a system of particles, the energy can depend on the higher powers of the field. Physically, this concerns systems with considerable many-particle interactions occurring along with the pair interaction. In the case of quadratic field-dependence, the partition function can be calculated exactly. However, exact calculation is impossible for more complicated dependences and we have to propose some approximate methods of partition function computation.

First, we consider the simplest local interaction with the energy depending only on the fourth power of the field [7, 15]. This dependence allows for the field self-action and may be obtained in the way as follows. The first method is to just restrict the general energy representation to the second term of the expansion in even powers of the field, i.e.,

$$E(\varphi) = \frac{1}{2} \int d\mathbf{r} d\mathbf{r}' \varphi(\mathbf{r}) V(\mathbf{r}, \mathbf{r}') \varphi(\mathbf{r}') + \frac{g}{4!} \int d\mathbf{r} \varphi^4(\mathbf{r}). \qquad (4.17)$$

In other words, we have to calculate the partition function Z of the form

$$\int D\varphi(\mathbf{r}) \exp\left\{-\frac{1}{T} E(\varphi(\mathbf{r}))\right\} \exp\left\{\frac{1}{T} \int d\mathbf{r} j(\mathbf{r}) \varphi^4(\mathbf{r})\right\}. \qquad (4.18)$$

We introduce new variables or simply change the scale of the variables, i.e., $\varphi \equiv \sqrt{T}\varphi$, $j \equiv \sqrt{T}j$, $g \equiv g/T$. Then the partition function Z in terms of new variables is given by

$$\int D\varphi \exp\left\{-\frac{1}{2} \int d\mathbf{r} d\mathbf{r}' \varphi V(\mathbf{r}, \mathbf{r}') \varphi' + \frac{g}{4!} \int d\mathbf{r} \varphi^4\right\}$$

$$\times \exp\left\{\frac{1}{T} \int j(\mathbf{r}) \varphi(\mathbf{r})\right\}. \qquad (4.19)$$

The formal calculation of the partition function is based on the expansion in power series of the small parameter g provided it is small indeed. Then the partition function may be written as

$$Z = \sum_{p=0}^{N} Z_p = \sum_{p=0}^{N} \frac{1}{p!} \left(\frac{g}{4!}\right)^p. \qquad (4.20)$$

For $j = 0$, expression (4.19) yields the free energy $F = -T \ln Z$. Sometimes, in the field theories, it is referred to as the vacuum contribution. Now the arbitrary correlation function is determined by the series in the parameter g, i.e.,

$$\chi^n(\mathbf{r}_1 \ldots \mathbf{r_n}) = \sum_{p=0}^{N} \frac{1}{p!} \left(\frac{g}{4!}\right)^p \frac{\delta}{\delta j(\mathbf{r}_1)} \cdots \frac{\delta}{\delta j(\mathbf{r_n})} Z_p. \qquad (4.21)$$

We employ the definition of the Gauss integral given above and then the partition function for the case under consideration is given by

$$Z = \left\{ -\frac{g}{4!} \int d\mathbf{r} \left(\frac{\delta}{\delta j(\mathbf{r})}\right)^4 \right\} \exp\left\{ \frac{1}{2} \int d\mathbf{r} d\mathbf{r'} j(\mathbf{r}) V^{-1}(\mathbf{r}, \mathbf{r'}) j(\mathbf{r'}) \right\}.$$
$$(4.22)$$

Thus we have general formulas for the partition function of a system whose points are described by some field. The field can provide information on the properties of the individual points of the system in the continuum case or characterize individual lattice sites in the discrete case. In the simplest case, the field may be associated with the coordinate. Let us consider the possibilities provided by the description of such a model of the local theory. If only pair interaction is present in the system, then its energy is described only by the first term of the general representation, i.e.,

$$E(\varphi) = \frac{1}{2} \int d\mathbf{r} d\mathbf{r'} \varphi(\mathbf{r}) V(\mathbf{r}, \mathbf{r'}) \varphi(\mathbf{r'}). \qquad (4.23)$$

We employ the Fourier transformation in a space of arbitrary dimension,

$$\varphi(x) = \int \frac{d^D k}{(2\pi)^m} \exp(ik \cdot x) \varphi(k),$$

and the expression for the pair interaction

$$V(x, x') = \int \frac{d^D k}{(2\pi)^D} \exp\left[ik \cdot (x - x')\right] V(k)$$

with $\int d^D x \exp(ik \cdot) x = (2\pi)^m \delta^D(k)$. Then we have

$$E(\varphi) = \frac{1}{2} \int \frac{d^m k}{(2\pi)^m} \varphi(-k) V(k) \varphi(k). \qquad (4.24)$$

In most cases $V(k)$ may be written in terms of the wave-vector series. This series provides a good description in the case of long-range potentials. It is less efficient in the case of short-range interactions, but the rescaling of the changes of the interaction character makes this expansion rather reasonable for short distances too. Moreover, in most cases the short-range interaction may be limited by the truncation length and hence the expansion is always correct for large distances. Thus in view of the specular symmetry, $V(k)$ may be written as a series in even powers of k, i.e., $V(k) = m^2 + ak^2 + bk^4 + \cdots$. The first term of this expansion is referred to as the mass term. It is associated with the energy for the case when the field is constant and the wave vector is equal to zero. The next terms are associated with the energy of changes caused by the field modification or, in other words, the energy of field "deformation". Such field modifications should be small because otherwise the expansion is invalid, i.e., this expansion corresponds to small changes of the field. Once the field changes are small, the expansion may be reduced to the two terms, i.e.,

$$E(k) = m^2 + k^2. \tag{4.25}$$

The second term of the expansion is written without coefficient since we can always perform a scaling transformation of the length or the wave vector which is actually the same. In the real space, we have

$$E = \frac{1}{2} \int dx \varphi(x)(-\partial^2 + m^2)\varphi(x). \tag{4.26}$$

Given that

$$\varphi \frac{\partial^2 \varphi}{\partial x^2} = \frac{\partial}{x}\left(\varphi \frac{\partial \varphi}{\partial x}\right) - \left(\frac{\partial \varphi}{\partial x}\right)^2,$$

the first term reduces to the integration over the surface where the field may always be taken equal to zero. Then we have only nonvanishing second term and the energy may be written in the form

$$E = \frac{1}{2} \int dx \left\{ \frac{1}{2}\left(\frac{\partial \varphi}{\partial x}\right)^2 + \frac{m^2}{2}\varphi^2 + \frac{g}{4!}\varphi^4 \right\}. \tag{4.27}$$

This is the well-known expression for the energy of a scalar field with self-action. It may be derived by another procedure making use of the same principle of integration over all probable field configurations or, to be more exact, over all probable fluctuations of the field. In the general case, we have some elastic medium described by an arbitrary field. Suppose we deal with a scalar field associated with the state of the system. In the ground state only this field is present, and this field is uniform. Naturally, fluctuations of arbitrary scales occur in the system as time passes, hence the field can become nonuniform. Thus, the action of this field should contain a term describing the probable spatially nonuniform distribution of the field. We may begin the calculation with the action of this scalar field. We assume, however, that the field interacts in some manner with the fluctuations of the medium that are not described by this field. In the thermodynamic description, this may concern the temperature or some other parameter. Suppose fluctuations of this additional field are random and their distribution is Gaussian. The latter assumption is not too strong to be invalid. In the case of independent fluctuations, the Gaussian distribution is the most realistic one. In order to consider the probable effect of fluctuations, we have to write the field action allowing for the interaction of the field with all probable fluctuations. We employ the path integral over all probable fluctuation configurations. The partition function is in this case given by

$$Z \sim \int D\varphi \int Dh \exp \left\{ - \int d \left(\mathbf{r} \left[\frac{1}{2} (\nabla \varphi)^2 - h\varphi - \frac{h^2}{D^2} \right] \right) \right\}, \quad (4.28)$$

where D^2 is the dispersion of fluctuations. The last term describes the energy of random fluctuations. Integration over all probable fluctuation configurations h yields an additional term in the action proportional to $m^2 \varphi^2$ with the coefficient $\mu^2 = D^2$. Thus, we have derived the mass term for the field. In order to find the term associated with the field self-action, it is sufficient to assume that the latter coefficient should have some average value plus fluctuations interacting with the field associated with the medium. We take the energy of these fluctuations into account and perform integration over all their probable configurations in the above manner. The averaging yields the energy contained in the action of the same form, i.e., $V(\varphi) = -(\mu^2/2\varphi^2) + (g/4!)\varphi^4$, where $\lambda \sim D*^2$ is proportional to the dispersion of new fluctuations. The averaging may be carried out under the assumption that the scales of fluctuations contributing to the mass coefficient and the nonlinearity factor differ, this provides both different values and different meaning of these coefficients.

Now we consider the expression for the potential. It is not difficult to note that if the function associated with the state of the system has the meaning, say, of the order parameter, then the expression for the energy completely reproduces the expression for the free energy corresponding to the description of phase transitions, except for the change of the sign of the mass coefficient before the critical temperature. And if, for example, the function has the meaning of coordinate, then the energy obtained reproduces the nonlinear oscillator energy. Thus we see that various systems can be described by a unified approach and the results obtained for some model formally are meaningful for other models too. This theory is local in the sense that energy values of the system depend on the field values of individual points.

In the local theory, it is suitable to write both the free energy and correlation functions by their Fourier transforms. We introduce the moment space and then the n-particle correlation function is given by

$$\chi^n(q_i) = \int \prod_{i=1}^{n} dq_i \exp(iq_i x_i) \chi^n(x_1 \ldots x_n).$$

We assume that the correlation function in the real space depends only on the difference of individual coordinates, thus we obtain a relation

$$\chi^n(q_i) = (2\pi)^D \delta^D(q_1 + \cdots + q_n) \chi^n(q_1 \ldots q_{n-1})$$

and the correlation function depending on two points is given by $\chi(q_1 q_2) = (2\pi)^D \delta^D(q_1 + q_2) \chi(q_1)$. Within the context of these observations, we find that $\int dx' V(x - x') \chi(x - x') = (-\partial_x + m^2) \chi(x - x') = \delta(x - x')$, then for the relevant Fourier transform, we have

$V(q)\chi(q) = 1$, and hence the correlation function is given by $\chi(q) = (1)/(q^2 + m^2)$.

4.2   Harmonic Lattice Model

Usually the model Hamiltonian has a harmonic addend in φ and there appears the Gaussian integral so that the partition trace may be calculated exactly. The starting point for the calculation of the Gaussian functional integral is the identity

$$\int_{-\infty}^{\infty} dx \exp\left\{-\frac{1}{2}ax^2 + bx\right\} = \left(\frac{2\pi}{a}\right)^{1/2} \exp\left\{\frac{b^2}{2a}\right\}. \qquad (4.29)$$

It is possible to derive the generalization of this result to the multidimensional integral. If we have A as $n \times n$ matrix with the components $A_{ij} = \langle i|A|j\rangle$ that is real and symmetric and may be diagonalized to obtain $\langle i|A|j\rangle = \delta_{ij}A_i$, then the Gaussian integral may be written as

$$\int_{-\infty}^{\infty} \left(\prod_i dx_i\right) \exp\left(-\frac{1}{2}x_i A_{ij} x_i + b_i x_i\right) = \prod_i \left(\frac{2\pi}{A_i}\right)^{1/2} \exp\left(\frac{b_i^2}{2A_i}\right).$$

$$(4.30)$$

The latter relation may be rewritten in another form, i.e.,

$$\prod_i \left(\frac{2\pi}{A_i}\right)^{1/2} \exp\left(\frac{b_i^2}{2A_i}\right) = \exp\left\{-\frac{1}{2}\text{Tr}\ln\frac{A}{2\pi} + \frac{1}{2}b_i A_{ij}^{-1} b_j\right\}.$$

$$(4.31)$$

This presentation may be applied to describe the harmonic lattice model. In this case, we may use the identity for multiple Gaussian integrals. The Hamiltonian of the harmonic lattice model is given by the standard expression

$$H_l = \frac{1}{2}\sum_{l,m} w_{l,m}\varphi_l\varphi_m = \frac{1}{2V}\sum_q w(q)|\varphi_q|^2, \qquad (4.32)$$

where

$$\varphi_l = \frac{1}{2v_0} \sum_q \varphi_q \exp\left(i\mathbf{q}\mathbf{R_l}\right), \quad w(q) = \frac{1}{v_0} \sum_q w_l \exp\left(i\mathbf{q}\mathbf{r_l}\right).$$

The eigenvalues of w_{lm} are $\langle q|w|q\rangle = v_0 w(\mathbf{q})$ and the free energy may be presented in the form

$$-T\ln Z = -T\ln\left\{\prod_l \int d\varphi_l \exp\beta\left(H + H^{\text{ext}}\right)\right\}, \quad (4.33)$$

where $H^{\text{ext}} = -\sum_l h_l \varphi_l$. For this sign of the external field, we obtain

$$-T\ln Z = -\frac{1}{2}T\sum_q \ln\left(\frac{\beta w(\mathbf{q})v_0}{2\pi}\right) - \frac{1}{2}\sum_{l,m} \beta h_l G^0_{l,m} h_m, \quad (4.34)$$

where

$$G^0_{l,m} = \int \frac{d^d q}{(2\pi)\mathbf{d}} \exp i\mathbf{q}(\mathbf{r_l} - \mathbf{r_m})\frac{kT}{w(\mathbf{q})} = \langle\varphi_l\varphi_m\rangle. \quad (4.35)$$

The free energy of the continuum model with

$$H_0 = \frac{1}{2}\int d^d x d^d x' w(\mathbf{x}, \mathbf{x}')\varphi(\mathbf{x})\varphi(\mathbf{x}') = \frac{1}{2V}\sum_q w(\mathbf{q})|\varphi_\mathbf{q}| \quad (4.36)$$

has the continuum limit

$$-T\ln Z = -\frac{1}{2}TV\int \frac{d^d q}{(2\pi)^d}$$

$$\times \ln\left(\frac{\beta w(\mathbf{q})v_0}{2\pi} - \frac{1}{2}\int d^d x d^d x' h(\mathbf{x})\beta G_0(\mathbf{x}, \mathbf{x}')h(\mathbf{x}')\right)$$

$$(4.37)$$

with $\beta w(\mathbf{q}) = G_0^{-1}(\mathbf{q})$ and

$$G_0(\mathbf{x}, \mathbf{x}') = \int \frac{d^d q}{(2\pi)^d} \exp\left[i\mathbf{q}(\mathbf{x} - \mathbf{x}')\right]\frac{kT}{w(\mathbf{q})} = \langle\varphi(\mathbf{x})\varphi(\mathbf{x}')\rangle. \quad (4.38)$$

The cell volume v_0 may be related to the cutoff Λ. The number of wave vectors $\mathbf{q}$ should be equal to the number of lattice points in the discrete theory. As a result, we have

$$\langle \varphi(\mathbf{q})\varphi(\mathbf{q}') \rangle = \delta_{q,q'} \frac{VkT}{w(\mathbf{q})}. \tag{4.39}$$

4.3   The n-Vector Model

The solution for the self-consistent field also exists in the $n \to \infty$ limiting case of the n-vector model [7, 15, 234]. We consider the n-vector model Hamiltonian as given by

$$H = \frac{1}{2} \int d^d x \left\{ a\varphi \cdot \varphi + k(\nabla\varphi)^2 + n^{-1} b(\varphi \cdot \varphi)^2 \right\}, \tag{4.40}$$

where $\varphi = (\varphi_1 \ldots \varphi_n)$ is the n-vector field and the quadratic term is explicitly proportional to n^{-1} and thus produces a meaningful $n \to \infty$ limit. We use for the Gaussian integration the relation

$$\exp\left\{ -\beta \frac{b(\varphi \cdot \varphi)^2}{n} \right\} = C \int_{-i\infty}^{i\infty} D\psi \exp\left\{ \frac{1}{16b} n\beta\psi^2 - \frac{1}{2}\beta(\varphi \cdot \varphi) \right\}, \tag{4.41}$$

where C is an unimportant constant. In this expression, the integration contour is along the imaginary rather than real axis because of the negative sign in the argument of the exponential on the right-hand side. The partition function may be rewritten as

$$Z = \int D\psi(\mathbf{x}) \exp\left\{ n\frac{\beta}{16b} \int d^d x \psi^2 - n\Phi(\psi(\mathbf{x})) \right\}, \tag{4.42}$$

where

$$\exp\left[-n\Phi(\psi(\mathbf{x})) \right] = \left\{ \int d\varphi(\mathbf{x}) \left[-\frac{\beta}{2} \int d^d x (a + \psi(\mathbf{x})\varphi(\mathbf{x}) \right. \right.$$
$$\left. \left. + k(\nabla\varphi)^2) \right] \right\}^n. \tag{4.43}$$

The integral on the right-hand side of this equation contains n identical replicas of the same integral and is the nth power integral over a scalar variable. This integral may be calculated using the previous result. Thus, we obtain

$$\Phi(\psi(\mathbf{x})) = \frac{1}{2} \text{Tr} \ln \left\{ \frac{\beta G^{-1}(\mathbf{x}, \mathbf{x}')}{2\pi} \right\}, \tag{4.44}$$

where

$$G^{-1}(\mathbf{x}, \mathbf{x}') = (a + \psi - k\nabla^2)\delta(\mathbf{x} - \mathbf{x}'). \tag{4.45}$$

The saddle-point approximation of the partition function is exact in the limiting case $n \to \infty$ and yields

$$\psi = 4b \int \frac{d^d q}{(2\pi)^d} \frac{kT}{a + \psi + kq^2} \tag{4.46}$$

and

$$TG^{-1}(\mathbf{q}) = a + \psi - k\nabla^2. \tag{4.47}$$

This is the statistical description of the n-vector model in the self-consistent approximation.

4.4   Potts Model

Hamiltonians of many lattice models may be expressed in terms of a sum over bond energies depending only on the differences of dynamic variables of pairs of sites that determine the bonds. This is possible if the bond potential may be Fourier-transformed and the partition function may be calculated either as a sum over original dynamic variables associated with lattice sites or as a sum over Fourier transform variables associated with the bonds. The transformation to the transform variables is called the duality transformation. The duality transformation also provides a direct relation between the lattice Coulomb gas and the discrete Gaussian model and is used to describe the transition from rough to smooth crystal structures [7, 15, 234]. The duality provides valuable insight into the phase transition in the two-dimensional model of interacting defects. Here, we consider the description of the duality transformation on a two-dimensional square lattice with bond energies of only nearest neighbor bonds. A square lattice with N sites has $2N$ bonds $\langle \mathbf{x}, \mathbf{x}' \rangle$ connecting nearest-neighbor sites $\mathbf{x}$ and $\mathbf{x}'$. The reduced Hamiltonian for this model is given by

$$\overline{H} = \frac{H}{kT} = \sum_{\mathbf{x}, \mathbf{x}'} V(\sigma_{\mathbf{x}} - \sigma_{\mathbf{x}'}), \tag{4.48}$$

where $\sigma_{\mathbf{x}}$ is the dynamical variable at the lattice site $\mathbf{x}$. The Hamiltonian for the N-state Potts model associates some energy with the nearest-neighbor bonds in the same state and different energy if they are in different states. The Hamiltonian of the Potts model may be written in the form:

$$H_{\text{Potts}} = -J \sum_{l,m} \left(N \delta_{\sigma_l, \sigma_m - 1} \right). \tag{4.49}$$

The three-state model may be made use of to describe the transition from the fluid to the structure of Kr absorbed on graphite. The two-state Potts model is the Ising model. The one-state model describes percolation. In the s-state of the Potts model $\sigma_{\mathbf{x}}$ is an integer-valued variable taking values $0, 1 \ldots, s-1$. Both $\sigma_{\mathbf{x}}$ and the difference $\sigma_l - \sigma_m$ are determined by modulo s so that they can take only similar values. $\exp[V(\sigma)]$ may be presented by the Fourier series

$$\exp[V(\sigma)] = \sum_{n=0}^{s-1} \exp\left\{ \frac{i2\pi n\sigma}{s} \right\} \exp[\widetilde{V}(n)] \tag{4.50}$$

or

$$\exp[\widetilde{V}(n)] = \frac{1}{s} \sum_{\sigma=0}^{s-1} \exp\left\{ -\frac{i2\pi n\sigma}{s} \right\} \exp[V(\sigma)]. \tag{4.51}$$

The Potts model partition function is given by

$$Z(V) = \sum_{\sigma} \prod_{\mathbf{x}, \mathbf{x}'} \exp[V(\sigma_{\mathbf{x}} - \sigma_{\mathbf{x}'})] \tag{4.52}$$

or

$$Z(V) = \frac{1}{s^N} \sum_\sigma \sum_{n_{x,x'}} \prod_{\mathbf{x},\mathbf{x'}} \exp[V(n_{x,x'})] \exp\left\{\frac{i2\pi n_{x,x'}(\sigma_\mathbf{x} - \sigma_\mathbf{x'})}{s}\right\}. \tag{4.53}$$

This partition function may be rewritten in the other form, i.e.,

$$Z(V) = \sum_{n_{x,x'}} \prod_\mathbf{x} \Delta(\mathbf{x}) \prod_{\mathbf{x},\mathbf{x'}} \exp\left\{V(n_{x,x'})\right\}, \tag{4.54}$$

where

$$\Delta(\mathbf{x}) = \frac{1}{s} \sum_\sigma \exp\left\{\frac{i2\pi\sigma \sum_a n_{x,x+a}}{s}\right\} \tag{4.55}$$

which is equal to one for $\sum_a n_{x,x+a} = 0$ and zero otherwise. The partition function is a sum over $2N$ bonds and function $\Delta(\mathbf{x})$ is definite at the site $\mathbf{x}$. This analysis is valid for different potential energies $V(\sigma)$. In the usual definition of the s-state of the Potts model we have

$$V(\sigma - \sigma') = sK\delta_{\sigma,\sigma'} = sK\delta_{\sigma-\sigma',0} \tag{4.56}$$

and

$$\exp[V(\sigma)] = 1 + [\exp(sK) - 1]\delta_{\sigma,0} \tag{4.57}$$

where $K = J/kT$ and the constant $-K$ is not included. With this form for $V(\sigma)$ we have

$$\exp[\widetilde{V}(n)] = \frac{1}{s}[\exp(sK) - 1]\exp(s\widetilde{K}\delta_{n,0}), \tag{4.58}$$

where

$$\exp(s\widetilde{K}) = \frac{s}{\exp(sK) - 1} \tag{4.59}$$

and the partition function is given by

$$Z(K) = \left\{\frac{1}{s}[\exp(sK) - 1]\right\}^N Z(\widetilde{K}). \tag{4.60}$$

If kT increases, then K decreases and $\widetilde{K}$ increases. If the Potts model contains the second-order phase transition for some $K = K_c$, then the partition function should be singular at K_c. If $Z(K)$ has a singularity, then there should occur a similar singularity at $\widetilde{K} = K_c$ and the equation for K_c should satisfy the condition

$$[\exp(sK) - 1]^2 = s. \tag{4.61}$$

Then the critical temperature is given by

$$kT_c = \frac{sJ}{\ln(1 + \sqrt{s})}. \tag{4.62}$$

The transition temperature for the Ising model or the two-state Potts model may be obtained in the form

$$kT_c = \frac{2J}{\ln(1 + \sqrt{2})}. \tag{4.63}$$

The partition function $Z(\widetilde{K})$ is equal to one when $\widetilde{K} = K = K_c$.

4.5   Villain and Gauss Lattice Models

In the xy-model the variable σ_x is transformed to the continuum angle variable θ_x and the potential $V(\theta)$ is a periodic function of θ with the period 2π. In the simplest version of the xy-model, we have [234]

$$V(\theta) = K(\cos\theta - 1). \tag{4.64}$$

The potential $V(\theta)$ should equal zero at its minimum for $\theta = 0$. The periodicity of the potential may be used in the Fourier presentation

$$\exp\left[V(\theta)\right] = \sum_{q=-\infty}^{\infty} \exp(in\theta)\exp[\widetilde{V}(q)], \tag{4.65}$$

where

$$\exp[\widetilde{V}(q)] = \frac{1}{2\pi}\int d\theta\,\exp(-iq\theta)\exp\left[V(\theta)\right]. \tag{4.66}$$

In our case of cosine potential we have

$$\exp\left[\widetilde{V}(q)\right] = \frac{1}{2\pi}\int d\theta \exp(-iq\theta)\exp\left\{K(\cos\theta) - 1\right\}$$
$$= \exp(-K)I_q(K), \tag{4.67}$$

where I_q is the modified Bessel function, and thus we have $\exp(\widetilde{V}(q)) = (K/2)^q/q!$ for $K \to 0$ and

$$\exp[\widetilde{V}(q)] = (2\pi K)^{-1/2}\exp\left(-\frac{q^2}{2K}\right)$$

for $K \to \infty$. The partition function for the xy-model may be transformed to dual variables similar to the previous case of the Potts model, i.e.,

$$Z(V) = \sum_{q_{x,x'}}\prod_{\mathbf{x}}\Delta(\mathbf{x})\prod_{\mathbf{x},\mathbf{x'}}\exp\left[V(q_{x,x'})\right]$$
$$= \sum_{q_R}\prod_{R,R'}\exp\left[\widetilde{V}(q_R - q_{R'})\right], \tag{4.68}$$

where $\mathbf{R}$ is a site on the dual lattice and $q_R = 0, \pm1, \pm2\ldots$ is an integer-valued function. For further transformations we employ the Poisson summation formula

$$\sum_{q=-\infty}^{\infty} g(q) = \sum_{m=-\infty}^{\infty}\int d\phi g(\phi)\exp(-i2\pi\phi m). \tag{4.69}$$

When applied to the periodic potential of the xy-model that may be written as a sum over integers q, we may write

$$\exp\left[V(\theta_{\mathbf{x}} - \theta_{\mathbf{x'}})\right] = \sum_{m_{x,x'}}\exp\left[V_0(\partial_x - \partial_{x'} - 2\pi m_{x,x'})\right], \tag{4.70}$$

where

$$\exp\left[V_0(\theta)\right] = \int_{-\infty}^{\infty} d\phi\exp\left[\widetilde{V}(\phi)\right]\exp(i\phi\theta). \tag{4.71}$$

Within the context of this expression the partition function takes the form

$$Z = \prod_x \int_0^{2\pi} d\theta_x \sum_{m_{x,x'}} \prod_{x,x'} \exp\left[V_0(\partial_x - \partial_{x'} - 2\pi m_{x,x'})\right]. \quad (4.72)$$

For low temperatures

$$\tilde{V}(q) \approx -\frac{q^2}{2K} - \frac{\ln(2\pi K)}{2}.$$

For this potential we have

$$\exp\left[V_0(\theta)\right] = \exp\left\{-\frac{K\theta^2}{2}\right\} \quad (4.73)$$

and the partition function is given by

$$Z = \prod_x \int_0^{2\pi} d\theta_x \sum_{m_{x,x'}} \prod_{x,x'} \exp\left\{-\frac{K}{2}(\partial_x - \partial_{x'} - 2\pi m_{x,x'})^2\right\}. \quad (4.74)$$

This is the Villain model. It is a version of the xy-model in which the periodic cosine potential is represented by the periodic function

$$\ln\left\{\sum_m \exp\left[-K\frac{(\theta - 2\pi m)^2}{2}\right]\right\}. \quad (4.75)$$

Finally, the Poisson summation formula may be applied to the dual version of the xy-model, then we obtain

$$Z = \sum_{q_R} \prod_{R,R'} (2\pi K)^{-1/2} \exp\left\{-\frac{1}{2K}(q_R - q_{R'})^2\right\} \quad (4.76)$$

that is the discrete Gaussian model. It is dual to the Villain version of the xy-model. Making use of the Poisson formula we obtain the partition function given by

$$Z = \sum_{q_R} \exp\left\{-2\pi^2 K \sum_{q_R, q_{R'}} q_R G(R - R')q_{R'}\right\}, \quad (4.77)$$

where $G(\mathbf{R})$ is the lattice Green function. The latter formula is final in the description of the lattice Coulomb gas.

4.6   Two-Dimensional Coulomb-Gas Models

In a system with continuum $U(1)$ symmetry, there can exist in the ordered phase a topologically stable defect that is referred to as vortex. Each defect of this type possesses the formation energy that consider the energy of the superfluid motion and the core where the value ψ of the order parameter tends to zero. With respect to this vortex, the fluctuation amplitude reduces to the average value of the order parameter. A sufficiently larger number of vortices may destroy the long-range order altogether and then thermally activated proliferation of the vortices leads to the phase transition to the high-temperature disordered phase. The value of the positive energy E_c is associated with the formation of the core of the vortex. The vortices in thermal equilibrium produce a term proportional to $\exp(-E_c/kT)$ in the partition function. The vortices reduce the spin-wave stiffness as well as the order-parameter amplitude. They produce new degrees of freedom of motion. It is possible to calculate the free energy of the xy-model with a vortex in which the phase $\theta(\mathbf{x})$ of the average order parameter $\langle\psi\rangle$ has uniform gradient. The partition function for a system in which $\theta(\mathbf{x})$ is spatially uniform may be obtained by taking into account the boundary condition that $\theta(\mathbf{x})$ should vanish at the edges of the sample [234]. Thus, even if $\theta(\mathbf{x})$ fluctuates, its volume-averaged gradient is equal to zero, i.e.,

$$\Omega^{-1}\int d^d x(\nabla\theta(\mathbf{x})) = \Omega^{-1}\int d\mathbf{S}\langle\theta(\mathbf{x})\rangle = 0, \qquad (4.78)$$

where the final integral extends over the surface of the sample and Ω is the volume of the sample. To calculate the energy associated with the spatially uniform gradient we take $\theta(\mathbf{x}) = \theta'(\mathbf{x}) + \mathbf{v}\cdot\mathbf{x}$ where θ' is constrained by the condition to vanish on the boundaries. Thus, we obtain

$$\nabla\theta = \Omega^{-1}\int d^d x\big[\nabla\theta(\mathbf{x}) + \mathbf{v}\big]\mathbf{v}. \qquad (4.79)$$

The free energy that describes the deformation of the uniform spatial distribution of θ is given by

$$F(\mathbf{v}) - F(0) = \Omega\rho_s(\mathbf{v})^2, \qquad (4.80)$$

where ρ_s is the macroscopic spin-wave stiffness. In a system with broken continuum symmetry in two spatial dimensions, the long-range order does not occur. The order parameter correlation function for low temperatures in a system with xy-symmetry possesses asymptotics differing from that of the temperature-dependence. As an example, the nonlinear model gives an indeterminate result for the transition temperature of the two-component xy-model in two dimensions. Topological point defects occur in the two dimensional xy-system but not in the Heisenberg system with $n \geq 3$. It is natural that thermally excited vortices might be responsible for the transition from the algebraic order to disorder in the two-dimensional xy-system. Kasterliz and Thouless [133] have proposed a simple heuristic argument that vortices can lead to a second-order phase transition in the xy-model and made a remarkably accurate estimate of the transition temperature. The energy of a simple vortex of unit strength in a sample of linear dimension R is given by [234]

$$E_v = \pi \rho_s \ln\left(\frac{R}{a}\right), \tag{4.81}$$

where a is a short-distance cutoff of the order of magnitude of the core radius. The core may be anywhere in the sample. The entropy $S = \ln(/Ra)$, and hence the free energy of the xy-system with a single vortex may be written in the form

$$F = E_v - kTS = (\pi \rho_s - 2kT)\ln\left(\frac{R}{a}\right). \tag{4.82}$$

When $kT < (\pi \rho_s)/2$, the minimum is not associated with a vortex. For the reversed inequality, the free energy has minima associated with vortices. Given that vortices destroy the phase, the temperature $kT_c = (\pi \rho_s)/2$ is the initial temperature that favors the production of vortices with the transition temperature from the ordered to disordered phase. The Green function in two dimensions is given by

$$G(\mathbf{x}) = \int \frac{d^2q}{(2\pi)^2} \frac{\exp(i\mathbf{qx})}{q^2} = \frac{1}{2\pi}\ln\frac{R}{a} - \frac{1}{2\pi}\ln\frac{|x|}{a} + C, \tag{4.83}$$

where C is constant. The first term in this equation diverges as the sample size increases. Thus, we have to introduce a scalar function that would present the vortex density and describe the states in a system where the total number of vortices is zero, the reason being the law of conservation of topological charges. In this case, the Hamiltonian of the system may be rewritten in the form [234]

$$H = H_{sw} + H_v, \qquad (4.84)$$

where

$$\frac{H_{sw}}{kT} = \frac{K}{2} \int d^2x (\nabla \theta)^2 \qquad (4.85)$$

is the spin-wave part of the reduced Hamiltonian where $K = \rho_s/kT$ and

$$\frac{H_v}{kT} = -\pi K \int d^2x d^2x' n_v(\mathbf{x}) n_v(\mathbf{x}') \ln \frac{|\mathbf{x} - \mathbf{x}'|}{a} + \frac{E_c}{kT} \sum_\alpha k_\alpha \qquad (4.86)$$

is the vortex part. The last term in the latter expression is the core energy of a defect with the topological charge k_α. The integral over positions contains a short-distance cutoff preventing two vortices from occupying the same position in the space. Since there is a minimum distance between vortices, it is often convenient to restrict this distance to the lattice constant a. The lattice version of the vortice Hamiltonian may be presented in the form

$$\frac{H_v}{kT} = -\pi K \sum_{l,m} k_l k_m \ln \frac{|\mathbf{R_l} - \mathbf{R_m}|}{a} + \frac{E_c}{kT} \sum_l k_l, \qquad (4.87)$$

where $\mathbf{R}_l$ are the lattice vectors. This presentation gives the Hamiltonian of the two-dimensional Coulomb gas with point charges k_l and charge density $n_v(\mathbf{x})$. The constraint $\sum_l k_l = 0$ is thus the constraint of charge neutrality. The vortex density correlation function may be obtained as [234]

$$\langle n_v(q) n_v(-q) \rangle = q^2 C_2, \qquad (4.88)$$

where

$$C_2 = -\frac{1}{4\Omega} \sum_{l,m} (\mathbf{R_l} - \mathbf{R_m})^2 \langle k_l k_m \rangle. \tag{4.89}$$

The charge correlation function in this equation is easily evaluated as power series. This model is very good for the study of phase transitions in the two-dimensional case. Its result is the occurrence of the second-order phase transition.

Chapter 5

Statistical Description of Condensed Matter

5.1 Partition Function for Model Systems

The unified approach makes it possible to describe equilibrium systems of interacting particles allowing for the formation of thermodynamically stable spatially inhomogeneous particle distributions and to consider the collective behavior of the structures formed. The topicality of this problem is confirmed by recent attempts made by other authors [7, 8, 16]. A wide range of systems of interacting particles occur, for which the statistics must be considered while dynamical quantum correlations may be disregarded. This means that the interaction may be treated in the classical manner. The Hamiltonian of such a system may be written as [53, 54, 68, 149, 150]

$$H(n) = \sum_s \varepsilon s n_s - \frac{1}{2} \sum_{s,s'} W_{ss'} n_s n_{s'} + \frac{1}{2} \sum_{s,s'} U_{ss'} n_s n_{s'}, \qquad (5.1)$$

where ε_s is the additive part of the particle energy in the state s (in most cases, it is the kinetic energy), $W_{ss'}$ and $U_{ss'}$ are the absolute values of the attraction and repulsion energies of particles in the states s and s', respectively. The macroscopic state of the system is determined by the occupation numbers n_s. The subscript s corresponds to the variables that describe individual particle states. It can also enumerate the lattice sites [68, 149]; the specifics of the lattice does not influence the result. Though the approach is adequate for the treatment of the discrete case as well, in this study we are interested only in the continuum approximation and assume that the medium is isotropic.

The partition function of the grand canonical ensemble $\sum_{\{n\}} \exp(-\beta H)$ is given by

$$Z = \sum_{\{n\}} \exp\left(-\beta H\right) \times \sum_{\{n\}} \exp\left\{-\beta\left[\sum_s \varepsilon_s n_s - \frac{1}{2}\sum_{s,s'} W_{ss'} n_s n_{s'}\right.\right.$$

$$\left.\left. + \frac{1}{2}\sum_{s,s'} U_{ss'} n_s n_{s'}\right]\right\}, \tag{5.2}$$

where $\sum_{\{n\}}$ implies summation over all probable distributions $\{n_s\}$, $\beta \equiv 1/kT$, and T is the absolute temperature. The summation in (5.2) can be formally carried out by introducing auxiliary field variables and making use of the known properties of the Gauss integrals [8, 10, 11, 13, 14], i.e.

$$\exp\left\{\frac{\nu^2}{2\theta}\sum_{s,s'} \omega_{ss'} n_s n_{s'}\right\} = \int_{-\infty}^{\infty} D\varphi \exp\left\{\nu \sum_s n_s \varphi_s\right.$$

$$\left. - \frac{\theta}{2}\sum_{s,s'} \omega_{ss'}^{-1} \varphi_s \varphi_{s'}\right\} \tag{5.3}$$

with

$$D\varphi = \frac{\prod_s d\varphi_s}{\sqrt{\det\left(2\pi\beta\omega_{ss'}\right)}}.$$

Here $\omega_{ss'}^{-1}$ is the inverse matrix that satisfies the equation $\omega_{ss''}^{-1}\omega_{s''s'} = \delta_{ss'}$ and $\nu^2 = \pm 1$ depending on the sign of the interaction or potential energy.

Within the context of (5.3), the partition function may be written as

$$Z = \int D\varphi \int D\psi \sum_{\{n_s\}} \exp\left\{\sum_s (\varphi_s + i\psi_s - \beta\varepsilon_s)\, n_s\right\} Q(\varphi, \psi),$$

$$\tag{5.4}$$

where

$$Q(\varphi, \psi) = \exp\left\{-\frac{1}{2\beta} \sum_{s,s'} \left(W_{ss'}^{-1} \varphi_s \varphi_{s'} + U_{ss'}^{-1} \psi_s \psi_{s'}\right)\right\}. \quad (5.5)$$

In the above analysis, we did not restrict the number of particles. Now let us fix the number of particles in the system, $N(n) = \sum_s n_s$. This means that we consider the canonical ensemble. To do this, we use the well known Cauchy formula, i.e.,

$$\frac{1}{2\pi i} \oint \xi^{(\sum_s n_s - N - 1)} d\xi = 1. \quad (5.6)$$

Then, making use of the contour integral [53, 54], we write the partition function of the system of N particles in terms of the grand partition function. Thus we may present Z_N as

$$Z_N = \frac{1}{2\pi i} \oint d\xi \int D\varphi \int D\psi Q(\varphi, \psi) \xi^{-(N+1)} \prod_s$$

$$\times \sum_{\{n_s\}} \left[\xi \exp\left(\varphi_s + i\psi_s - \beta\varepsilon_s\right)\right]^{n_s}. \quad (5.7)$$

After summing over the occupation numbers n_s, the partition function reduces to

$$Z_N = \frac{1}{2\pi i} \oint d\xi \int D\varphi \int D\psi \, e^{-S(\varphi, \psi, \xi)}, \quad (5.8)$$

where

$$S(\varphi, \psi, \xi) = \frac{1}{2\beta} \sum_{s,s'} \left(W_{ss'}^{-1} \varphi_s \varphi_{s'} + U_{ss'}^{-1} \psi_s \psi_{s'}\right)$$

$$+ \delta \sum_s \ln\left(1 - \delta\xi e^{-\beta\varepsilon_s + \varphi_s} \cos\psi_s\right) + (N+1)\ln\xi. \quad (5.9)$$

In the latter expression, the sign $\delta = \pm 1$ depends on the statistics (plus and minus for the Bose and Fermi systems, respectively).

The representation of the partition function by the functional integral over auxiliary fields corresponds to the construction of an equilibrium sequence of probable states of the system as per their weights. This representation provides a possibility to employ the well-known methods of quantum field theory and to avoid using the perturbation theory. The extension to the complex plane makes it possible to apply the saddle-point method [9].

It should be emphasized that this description is useful [7, 8] because it advances the study of thermodynamic characteristics of model systems and their dependence on the medium. If interaction of some type does not occur, then the general representation (5.8) and (5.9) makes it possible to reproduce the partition function for pure gravitational attraction given in [8], and the partition function of the sine-Gordon model for pure Coulomb repulsion given in [7]. This approach also makes it possible to consider the states with spatially inhomogeneous particle distributions. To do this, one has to vary the functional $S(\varphi, \psi, \xi)$ for the fields φ and ψ and the analog of the chemical potential ξ, and then to apply the saddle-point method in order to obtain the asymptotic value of the partition function Z_N for $N \to \infty$. The solutions, in which the action S remains finite when the volume of the system tends to infinity, may be interpreted as thermodynamically stable particle distributions. Whether the distribution is homogeneous or inhomogeneous, depends on the solutions that satisfy the extremum condition for the functional S, i.e., $\delta S/\delta\varphi = \delta S/\delta\psi = \delta S/\delta\xi = 0$. The equations for the saddle-point states are given by

$$\frac{1}{\beta} \sum_{s'} W_{ss'}^{-1} \varphi_{s'} - \frac{\xi_s e^{\varphi_s} \cos\psi_s}{1 - \delta\xi_s e^{\varphi_s} \cos\psi_s} = 0$$

$$\frac{1}{\beta} \sum_{s'} U_{ss'}^{-1} \psi_{s'} + \frac{\xi_s e^{\varphi_s} \sin\psi_s}{1 - \delta\xi_s e^{\varphi_s} \cos\psi_s} = 0 \qquad (5.10)$$

$$\sum_{s} \frac{\xi_s e^{\varphi_s} \cos\psi_s}{1 - \delta\xi_s e^{\varphi_s} \cos\psi_s} = N + 1,$$

where $\xi_s = \xi e^{\beta\varepsilon_s} = e^{\beta(\varepsilon_s - \mu)}$, and μ is the chemical potential.

This set of equations provides a solution of the above many-particle problem in the sense that it selects the system states whose contributions to the partition function are dominant. As in the general case the inverse of the interaction matrix is not defined uniquely, it is impossible to find the general solution of the set (5.10). And even if the latter problem can be solved, there still remain technical difficulties associated with the need to solve a set of nonlinear equations and to interpret the solutions thereof. In the sections that follow we consider the solutions for some model systems.

It should be noted that within the context of the normalization condition (the third equation in the set (5.10)) we may regard the expression

$$f_s = \frac{\xi_s e^{\varphi_s} \cos \psi_s}{1 - \delta \xi_s e^{\varphi_s} \cos \psi_s}$$

as the particle distribution function determined by the auxiliary fields. It is obvious that, for given statistics, the distribution function depends on the nature and intensity of the interaction. Moreover, the proposed representation may be used to extend the treatment of the Bose condensation to the coordinate space. The cluster formation corresponds to particle localization within a limited space. In our treatment, the effect is manifest in the behavior of the auxiliary fields and chemical potential. Probably, the proposed approach will improve the understanding of the fractional statistics of particles too [8, 151, 152].

In what follows we apply the approach to describe model systems with various interactions, for which the partition function may be estimated in the thermodynamic limiting case. With this goal, we make use of the continuum approximation that makes it possible to obtain analytic expressions even for inhomogeneous particle distributions. In the continuum approximation, the subscript s runs through a continuum of values in the system volume V. When integrating over moments and coordinates, we bear in mind that the unit cell volume in the space of individual states is equal to $\omega = (2\pi\hbar)^3$.

In the continuum case, the inverse matrix $w_{ss'}^{-1}$ for the interaction $w_{ss'} = w(|\mathbf{r}_s - \mathbf{r}_{s'}|)$ is given by [8, 10, 14]

$$w_{rr'}^{-1} = \delta_{rr'} \hat{L}_{r'}, \tag{5.11}$$

where $\hat{L}_{r'}$ is the operator for which the interaction potential is the Green function. For the screened Coulomb or Newtonian potential, the inverse operator may be written as [10, 11, 13, 14]

$$\hat{L}_{r'} = -\frac{1}{4\pi q^2}\left(\Delta_{r'} - \lambda^2\right), \tag{5.12}$$

where q^2 is the interaction constant and λ^{-1} is the screening length. The number of realistic interactions, for which the inverse operator can be found, is limited. The difficulties in obtaining the inverse operator can be avoided by introducing a collective variable [23, 153] that corresponds to the relationship between the introduced fields on the saddle-point trajectory. In this approach, it is easier to find the required parameters of inhomogeneous structures. It is difficult, however, to trace the formation details for such particle distributions. Without loss of generality within the above approximations, in what follows, we describe the model systems of interacting particles taking into account their spatially inhomogeneous distributions. First of all, we demonstrate the advantage of the approach for the well-known models, and then describe the actual systems with interaction and find the conditions for the cluster formation and cluster parameters. To demonstrate the advantages of the approach, we first derive the well-known results.

5.2 Ideal Classical and Quantum Gases

For the Boltzmann statistics in the continuum case, (5.8) reduces to

$$Z_N = \frac{1}{2\pi i}\oint d\xi \exp\left\{\frac{1}{\omega}\int dV \int d^3p\left[\xi e^{-\beta\varepsilon} - (N+1)\ln\xi\right]\right\}. \tag{5.13}$$

Since $\varepsilon = p^2/2m$, where m is the particle mass, then

$$\int d^3p\, e^{-(p^2/2m)\beta} = \left(\frac{2\pi m}{\beta}\right)^{3/2}$$

and hence

$$Z_N = \frac{1}{2\pi i} \oint d\xi \exp\left\{ \xi V \left(\frac{2\pi m}{\beta \hbar^2} \right)^{3/2} - (N+1)\ln \xi \right\}$$

$$= \frac{1}{2\pi i} \oint d\xi \, e^{-S(\xi)}. \tag{5.14}$$

We find the saddle-point value $\tilde{\xi}$ from the equation $\delta S / \delta \xi = 0$ and thus obtain

$$\tilde{\xi} = \left(\frac{2\pi m}{\beta \hbar^2} \right)^{-3/2} \frac{N+1}{V}.$$

Now we substitute this value in (5.14) and apply the Stirling formula $N - N \ln N \simeq \ln N!$. This yields the partition function of the Boltzmann gas as

$$Z_N^0 = \frac{V^{N+1}}{(N+1)!} \left(\frac{2\pi m k T}{\hbar^2} \right)^{3(N+1)/2}. \tag{5.15}$$

These well-known results for the ideal systems show the consistency of the approach proposed here with the traditional methods. Later on, we demonstrate new advantages of this statistical description.

In the case of an ideal quantum gas, $\varphi = \psi = 0$ and the partition function (5.8) may be written as

$$Z_N = \frac{1}{2\pi i} \oint d\xi \exp\left\{ -\delta \sum_s \ln \left(1 - \delta \xi e^{-\beta \varepsilon_s} \right) - (N+1)\ln \xi \right\}$$

$$= \frac{1}{2\pi i} \oint \frac{d\xi}{\xi^{N+1}} \prod_s \left(1 - \delta \xi e^{-\beta \varepsilon_s} \right)^\delta. \tag{5.16}$$

Since the partition function of the grand canonical ensemble is [53, 54] $Z = \sum_N \xi^N Z_N$, we thus obtain the known result

$$Z = \prod_s \left(1 - \delta \xi e^{-\beta \varepsilon_s} \right)^\delta. \tag{5.17}$$

Let us derive some expressions used in this section for ideal Fermi and Bose gases. Thus, we shall demonstrate the correctness and greater rationality of our approach as compared to the traditional method [53].

For Fermi gas, we have $\varphi = \psi = 0$. Then the action for the system is given by

$$
S = -\frac{1}{\omega} \int dV \int 4\pi p^2 dp \ln\left(1 + \xi \exp\left(-\beta \varepsilon_p\right)\right) + (N+1)\ln\xi
$$
$$
= -\frac{V}{\lambda^3} f_{5/2}\left(\xi\right) + (N+1)\ln\xi, \tag{5.18}
$$

where

$$
f_{5/2}\left(\xi\right) = \frac{4}{\sqrt{\pi}} \int_0^\infty dx x^2 \ln\left(1 + \xi e^{-x^2}\right) = \sum_{l=1}^\infty (-1)^{l+1} \frac{\xi^l}{l^{5/2}}
$$

is a special Fermi function [53, 54]; $\varepsilon_p = p^2/2m$ is the kinetic energy of a particle; $\lambda = \sqrt{(\beta\hbar^2)/(2\pi m)}$ is the thermal wavelength of a particle. Then the equation for the saddle point is given by

$$
\frac{1}{v} = \frac{1}{\lambda^3} f_{3/2}\left(\xi\right), \quad f_{3/2}\left(\xi\right) = \xi \frac{\partial f_{5/2}\left(\xi\right)}{\partial \xi}, \quad v \equiv \frac{V}{N}. \tag{5.19}
$$

The partition function for this case is given by

$$
Z_N = \exp\left\{\frac{V}{\lambda^3} f_{5/2}\left(\xi\right) - (N+1)\ln\xi\right\}, \tag{5.20}
$$

where ξ is determined by Eq. (5.19). The knowledge of the partition function makes it possible to find any thermodynamical functions. Let us consider the case $\xi \to 0$ corresponding to high temperatures (the Boltzmann limiting case). Under this condition, Eqs. (5.19) and (5.20) reduce to

$$
Z_N = \exp\left\{\frac{V}{\lambda^3}\xi - (N+1)\ln\xi\right\} \approx \frac{V^N}{N!}\left(\frac{mkT}{2\pi\hbar^2}\right)^{3/2N} \tag{5.21}
$$

$$
\frac{1}{v} = \frac{\xi}{\lambda^3}. \tag{5.22}
$$

Now, we consider the case $T \to 0$ corresponding to the degenerated Fermi gas. Then the special Fermi functions are given by [53]

$$f_{3/2} = \frac{4}{3\sqrt{\pi}} \left[(\ln \xi)^{3/2} + \frac{\pi^2}{8} (\ln \xi)^{-1/2} \right] + O(\xi^{-1})$$

$$f_{5/2} = \frac{4}{3\sqrt{\pi}} \left[\frac{2}{5} (\ln \xi)^{5/2} + \frac{\pi^2}{4} (\ln \xi)^{1/2} \right] + O(\xi^{-1}). \tag{5.23}$$

Then, the expression (5.19) is reduced to

$$\frac{\lambda^3}{\upsilon} \approx \frac{4}{3\sqrt{\pi}} (\ln \xi)^{3/2} \Rightarrow \xi \approx e^{\beta \varepsilon_F}, \tag{5.24}$$

where

$$\varepsilon_F \equiv \frac{\hbar^2}{2m} \left(\frac{6\pi^2}{\upsilon} \right)^{2/3} \tag{5.25}$$

is the Fermi energy.

We can find analogical expressions for the Bose gas. We have to start from the action

$$S = \frac{1}{\omega} \int dV \int 4\pi p^2 dp \ln \left(1 - \xi \exp \left(-\beta \varepsilon_p \right) \right) + (N+1) \ln \xi$$

$$= -\frac{V}{\lambda^3} g_{5/2} \left(\xi \right) + \ln \left(1 - \xi \right) + (N+1) \ln \xi, \tag{5.26}$$

where

$$g_{5/2} \left(\xi \right) = -\frac{4}{\sqrt{\pi}} \int_0^\infty dx x^2 \ln \left(1 - \xi e^{-x^2} \right) = \sum_{l=1}^{\infty} \frac{\xi^l}{l^{5/2}}$$

is the special Bose function [53, 54]. It is clear that the chemical activity is always less than one, $\xi < 1$, unlike the Fermi system; $\ln(1 - \xi)$ is the action for the condensed phase (the addendum with $\mathbf{p} = 0$ is as important as the rest of the sum when $\xi \to 1$).

Now we derive some expressions used in this section for the ideal Bose gas. In this case $\varphi = \psi = 0$. Then the action for the system is

given by

$$S = \frac{1}{\omega} \int dV \int 4\pi p^2 dp \ln\left(1 - \xi \exp\left(-\beta\varepsilon_p\right)\right) + (N+1)\ln\xi$$

$$= -\frac{V}{\lambda^3}g_{5/2}\left(\xi\right) + \ln\left(1 - \xi\right) + (N+1)\ln\xi, \qquad (5.27)$$

where

$$g_{5/2}\left(\xi\right) = -\frac{4}{\sqrt{\pi}}\int_0^\infty dx\, x^2 \ln\left(1 - \xi e^{-x^2}\right) = \sum_{l=1}^\infty \frac{\xi^l}{l^{5/2}}.$$

Hence, it is clear that the activity is always less than one, $\xi < 1$, unlike the Fermi system; $\varepsilon_p = p^2/2m$ is the kinetic energy of a particle; $\ln\left(1 - \xi\right)$ is the action for the condensed phase (the addend with $\mathbf{p} = 0$ that is as important as the rest of the sum when $\xi \to 1$); $\lambda = \sqrt{(\beta\hbar^2)/2\pi m}$ is the wave thermal length of a particle. Then the equation for the saddle point is given by

$$\frac{1}{v} = \frac{1}{\lambda^3}g_{3/2}\left(\xi\right) + \frac{1}{V}\frac{\xi}{1 - \xi} \iff \frac{\langle n_0\rangle}{V} = \frac{\lambda^3}{v} - g_{3/2}\left(\xi\right), \qquad (5.28)$$

where

$$g_{3/2}\left(\xi\right) = \xi\frac{\partial g_{5/2}\left(\xi\right)}{\partial\xi}, \qquad v \equiv \frac{V}{N},$$

$\langle n_0\rangle$ is at zero level. Then the value $\langle n_0\rangle/V$ is positive under the condition

$$\frac{\lambda^3}{v} > g_{3/2}\left(1\right). \qquad (5.29)$$

Thus, we have a condensation of a finite number of particles on the level with $p = 0$ (the so-called Bose–Einstein condensation). The portion of condensed particles is determined by the expression (5.28):

$$\frac{\langle n_0\rangle}{N} = 1 - \frac{v}{v_c} = 1 - \left(\frac{T}{T_c}\right)^{3/2}, \qquad (5.30)$$

where the critical parameters are given by

$$v_c = \frac{\lambda^3}{g_{3/2}(1)}, \quad T_c = \frac{2\pi\hbar^2}{[vg_{3/2}(1)]^{(2/3)}mk}. \tag{5.31}$$

Now, we consider the case $\xi < 1$. Being at zero level is as insignificant as at other levels with $p \neq 0$, which means that the condensed phase does not occur. Such situation is realized under the conditions $T > T_c$ and $v > v_c$. The partition function for this case is given by

$$Z_N = \exp\left\{\frac{V}{\lambda^3}g_{5/2}(\xi) - (N+1)\ln\xi\right\}, \tag{5.32}$$

where ξ is determined by Eq. (5.28) under the condition that there is no condensed phase. The thermodynamical functions may be found once the partition function is known. The free energy and the pressure of the system are given by

$$F = -kT\frac{V}{\lambda^3}g_{5/2}(\xi) + NkT\ln\xi, \tag{5.33}$$

$$P = kT\frac{1}{\lambda^3}g_{5/2}(\xi). \tag{5.34}$$

The next case to be considered is $\xi \longrightarrow 0$. It corresponds to high temperatures (the Boltzmann limiting case). Under this condition, expressions (5.28) and (5.32) are reduced to

$$Z_N = \exp\left\{\frac{V}{\lambda^3}\xi - (N+1)\ln\xi\right\} \approx \frac{V^N}{N!}\left(\frac{mkT}{2\pi\hbar^2}\right)^{3/2}N, \tag{5.35}$$

$$\frac{1}{v} = \frac{\xi}{\lambda^3}. \tag{5.36}$$

Finally, we consider the case $\xi \to 1$. In this case $\langle n_0\rangle/V$ is finite and is determined by formula (5.30). This means that the condensed phase occurs in the system. Such situation is realized for sufficiently low temperatures and small volumes, $(T < T_c, v < v_c)$.

The expression (5.28) is reduced to

$$\frac{\xi}{1-\xi} \gg \frac{V}{\lambda^3} g_{3/2}(1) \Rightarrow N + 1 \approx \frac{\xi}{1-\xi}. \tag{5.37}$$

In order to find the free energy for this case, we define the partition function

$$Z_N = \exp\left\{\frac{V}{\lambda^3} g_{5/2}(1) - \ln(1 - \xi)\right\}, \tag{5.38}$$

then

$$\frac{F}{NkT} = -\frac{v}{\lambda^3} g_{5/2}(1), \tag{5.39}$$

$$P = kT \frac{1}{\lambda^3} g_{5/2}(1). \tag{5.40}$$

Now we shall find the internal energy U of the system proceeding from the predetermined analogy between the technique of thermodynamics in our representation and the field theory. In order to do it, we use the correlation $-H = \partial S_{\text{mech}}/\partial t$ and determine $H \longleftrightarrow U, S_{\text{mech}} \longleftrightarrow S_{\text{term}}, t \longleftrightarrow 1/kT$, where H is the Hamiltonian of the system, S_{mech} and S_{term} are the actions for the mechanical and thermodynamical (5.27) systems, respectively, t is time, and $1/kT$ is the reciprocal temperature. Then, making use of the expression (5.27), we have

$$U = -\frac{\partial S}{\partial(1/kT)} = \frac{3}{2} \frac{VkT}{\lambda^3} g_{5/2}(\xi) = \frac{3}{2} PV. \tag{5.41}$$

Expressions (5.33)–(5.35), (5.39)–(5.41) reproduce the formulas obtained by the usual methods [53, 54]. This fact confirms the correctness of the approach proposed.

5.3 Hard Spheres Model

We generalize the hard spheres model by assuming that the potential barrier U_0 is finite. This value is determined by the mechanism of particle collisions. The interaction energy may be written as $U_{ss'} = U_0 \delta_{ss'}$. In this case, the inverse operator is described by the expression

$U_{ss'}^{-1} = U_0^{-1} \delta_{ss'}$. In the continuum case, we can approximately invert the potential. When doing this, we have to remove the self-interaction terms that arise as we pass from the discrete sum $\sum_{s,s'} U_{ss'} n_s n_{s'}$ to continuum. Thus, (5.8) and (5.9) reduce to

$$Z_N = \frac{1}{2\pi i} \oint d\xi \int D\psi \, e^{-S(\psi,\xi)} \qquad (5.42)$$

and

$$S(\psi,\xi) = \int dV \left[\frac{1}{2\beta U_0 V} \psi^2 - \xi A \cos \psi \right] + (N+1) \ln \xi, \qquad (5.43)$$

where $A = (2\pi m/\beta\hbar^2)^{3/2}$. The saddle-point equations that satisfy the extremum condition for the action S are given by

$$\frac{1}{\beta U_0 V} \psi + \xi A \sin \psi = 0$$

$$\int dV \xi A \cos \psi = N+1. \qquad (5.44)$$

From the behavior of the interaction potential ($U = U_0$ for $r < r_0$, and $U = 0$ for $r > r_0$, where r_0 is the particle radius) we conclude that $\psi = 0$ everywhere except for the volume $V_0 = 2v_0(N+1)$ (v_0 is the particle volume), and $\psi = \tilde\psi$ in the volume V_0 where particle interaction occurs. The quantity $\tilde\psi$ may be found from the equation

$$\tilde\psi + \xi A \beta U_0 V \sin \tilde\psi = 0 \qquad (5.45)$$

with the normalization condition

$$\xi A (V - V_0) + \xi A V_0 \cos \tilde\psi = N+1. \qquad (5.46)$$

If the interaction energy is finite, then the solution of Eq. (5.45) may be written in the form $\tilde\psi \simeq \pi + \alpha$, $\alpha \ll \pi$, and we find that $\alpha = \pi/\xi A \beta U_0 V$. As $\beta U_0 \to \infty$, we have $\tilde\psi = \pi$. This corresponds to the pure hard spheres model. Having applied the successive approximations to calculate the chemical potential, we rewrite Eq. (5.46) in the form

$$\xi A (V - V_0) - \xi A V_0 \simeq N + 1,$$

then

$$\xi \simeq \frac{N+1}{AV} \left(1 - \frac{2V_0}{V}\right),$$

and thus

$$\tilde{\psi} \simeq \pi \frac{(N+1)\beta U_0}{(N+1)\beta U_0 - (1 - 2V_0/V)}.$$

The solutions obtained make it possible to estimate the action. We have

$$S = \frac{\pi^2}{\beta U_0} \frac{V_0}{V} \left[\frac{(N+1)\beta U_0}{(N+1)\beta U_0 - (1 - 2V_0/V)}\right]^2$$

$$- (N+1) + (N+1) \ln\left(\frac{N+1}{AV(1 - 2V_0/V)}\right) \qquad (5.47)$$

and thus the partition function is given by

$$Z_N = \frac{V^{N+1}}{(N+1)!} \left(\frac{2\pi mkT}{\hbar^2}\right)^{3(N+1)/2} \left(1 - \frac{2V_0}{V}\right)^{N+1}$$

$$\times \exp\left\{-\frac{\pi^2}{\beta U_0} \frac{V_0}{V} \left[\frac{(N+1)\beta U_0}{(N+1)\beta U_0 - (1 - 2V_0/V)}\right]^2\right\}. \qquad (5.48)$$

The latter expression may be rewritten in a compact form

$$Z_N = Z_N^0 \left(1 - \frac{2V_0}{V}\right)^{N+1} \left\{1 - \frac{\pi^2}{\beta U_0} \frac{V_0}{V}\right.$$

$$\times \left.\left[\frac{(N+1)\beta U_0}{(N+1)\beta U_0 - (1 - 2V_0/V)}\right]^2\right\}, \qquad (5.49)$$

where Z_N^0 is the partition function of the ideal Boltzmann gas (5.15). For $U_0 \to \infty$, we find the partition function of the pure hard spheres model to be of the form

$$Z_N = Z_N^0 \left(1 - \frac{4v_0 \, (N+1)}{V} \right)^{N+1}. \tag{5.50}$$

This result may be immediately obtained from the solution of Eq. (5.44) under the assumption $U_0 \to \infty$. It should be emphasized that solution (5.50) exactly reproduces the partition function of the hard spheres model. In this approach, it is derived without calculating virial coefficients.

Hard spheres model with short-range attraction

Now we generalize the hard spheres model by assuming that the potential barrier U_0 is infinite and there exists a short-range attractive potential $W = \text{const}$ with the range R. The interaction energy may be written as $W_{ss'} = W_0 \theta_{ss'}$. In this case, the inverse operator is described by the expression $U_{ss'}^{-1} = W^{-1}$ within the region from the sphere radius to the distance R. In the continuum case, we can approximately invert the potential. When doing this, we have to remove the self-interaction terms that arise as we pass from the discrete sum $\sum_{s,s'} W_{ss'} n_s n_{s'}$ to continuum. Thus, (5.8) and (5.9) reduce to

$$Z_N = \frac{1}{2\pi i} \oint d\xi \int D\varphi \int D\psi \, e^{-S(\varphi\psi,\xi)}, \tag{5.51}$$

but the effective action consists of three addends for different areas. The first term of the free energy is associated with the volume of the hard sphere, where only repulsion with infinite potential occurs. In this region $\varphi = 0$ and

$$S_1(\psi, \varphi, \xi) = - \int_0^{V_0} dV \xi A \cos \psi. \tag{5.52}$$

The saddle-point solution is presented as $\sin \psi = 0$ or $\psi = 0, \pi,$ $2\pi, \ldots$. For a hard sphere, similarly to the previous case, the solution may be taken of the form $\psi = \pi$ and $S_1 = \xi A V_0$. For the second region, from the sphere radius to distance R, where attractive interaction occurs, we may write

$$S_2(\psi, \varphi, \xi) = \int_{V_0}^{V_R} dV \left[\frac{1}{2\beta W V} \varphi^2 - \xi A e^\varphi \right]. \tag{5.53}$$

For this case we may write the Euler–Lagrange equation in the form

$$\frac{1}{\beta W V} \varphi - \xi A e^\varphi = 0 \tag{5.54}$$

and the effective action may be presented as

$$S_2(\psi, \varphi, \xi) = \frac{1}{\beta W V} \left(\frac{1}{2} \varphi_0 - 1 \right) \varphi_0,$$

where φ_0 is the solution of the previous equation. In the region larger than the range of action of the attractive interaction $\varphi = 0$ and $\psi = 0$, and the effective action may be presented as $S_3(\psi, \varphi, \xi) = \xi A(V - V_R)$, where V is the volume of the system. The effective action for the system may be written in the form

$$S_1(\psi, \varphi, \xi) = -\xi A(V - V_R - V_0)$$
$$+ \frac{1}{\beta W V} \left(\frac{1}{2} \varphi_0 - 1 \right) \varphi_0 + (N+1) \ln \xi. \tag{5.55}$$

The solution for φ_0 may be written as $\varphi_0 = W(-\alpha)$, where $W(-\alpha)$ is the product of log function of the argument $\alpha = \xi A \beta W(V_R - V_0)$. For small α we may use $\varphi = -\alpha$ and present the effective action in the form

$$S_1(\psi, \varphi, \xi) = -\xi A(V - 2V_0) + \frac{1}{2} (\xi A)^2 \beta W(V_R - V_0) + (N+1) \ln \xi. \tag{5.56}$$

From the normalization condition, we obtain the equilibrium value of ξ as

$$\xi A (V - 2V_0) + \frac{1}{2} (\xi A)^2 \beta W(V_R - V_0) = N + 1. \tag{5.57}$$

If the interaction energy βW is small, we can apply the method of consistent approximations. The first step is to find the chemical activity through the condition

$$\xi A \left(V - 2V_0 \right) - \xi A V_0 \simeq N + 1$$

and to substitute this value to the effective action. Then, we may present the partition function in a simple form as

$$Z_N = \frac{V^{N+1}}{(N+1)!} \left(\frac{2\pi m k T}{\hbar^2} \right)^{3(N+1)/2} \left(1 - \frac{2V_0}{V} \right)^{N+1}$$
$$\times \exp \left\{ -\frac{1}{2} \left(\frac{N+1}{V - 2V_0} \right)^2 \beta W \left(V_R - V_0 \right) \right\}. \tag{5.58}$$

The latter expression may be rewritten in a compact form

$$Z_N = Z_N^0 \left(1 - \frac{4v_0 \left(N + 1 \right)}{V} \right)^{N+1}$$
$$\times \exp \left\{ -\frac{1}{2} \left(\frac{N+1}{V - 2V_0} \right)^2 \beta W \left(V_R - V_0 \right) \right\}, \tag{5.59}$$

where Z_N^0 is the partition function of the ideal Boltzmann gas (5.15). As before, this partition is derived without calculating the virial coefficient.

5.4 Two Exactly Solvable Models of Statistical Physics

Only a few model systems of interacting particles are known in statistical physics [149] for which exact solutions have been found for the thermodynamic limiting case. For the most realistic Ising model, the solution is obtained only for one and two dimensions. To find the exact solutions for the three-dimensional models is a more complicated problem, though predictions based on the universality principles are confirmed experimentally [53, 54]. For the time being, no exact solution has been found for the three-dimensional models with short-range forces. In [2] a three-dimensional lattice model is proposed, for which the partition function is presented in terms of known functions and an exact solution is found in this sense. This model deals with a system of particles interacting in a lattice with an unlimited number of particles at each site. The partition function is calculated and the lattice effect on thermodynamic parameters of the system is considered for the case when particles from different sites attract and particles from the same site repel. In this section, another exactly solvable three-dimensional model of a system of particles interacting in a lattice is proposed. As before, the model is based on the assumption that the number of particles at each site is not limited. Particles from different sites are repelled and particles from the same site are attracted. The approach of [2] in constructing the generating partition function makes it possible to consider two model systems with opposite interaction mechanisms in a unified manner, and to calculate the partition function. In the continuum limiting case, the equation of state can be derived and phase transition conditions can be discussed. Let us consider a model system of interacting particles in a lattice with the microstates described by a set of site occupation numbers n_s. The configuration Hamiltonian of a system with the two above-mentioned opposite interaction types may be written in a unique way, i.e.,

$$H(n) = \sum_s \varepsilon_s n_s - \nu^2 \frac{1}{2} \sum_{s,s'} W_{ss'} n_s n_{s'} + \nu^2 \frac{1}{2} \sum_s U_s n_s^2, \qquad (5.60)$$

where $\nu^2 = \pm 1$ depending on the sign of the interaction or the potential energy. The upper sign corresponds to the model with attraction at different sites and repulsion at the same site, the lower one is associated with the contrary situation — repulsion at different sites and attraction at the same site of the initial lattice. The quantities $W_{s,s'}$ and $U_{s,s'}$ describe the spatial dependence of relevant interaction potentials. It should be noted that the effective Hamiltonian of the Gauss model of spin behavior in fractals may be reduced to the same form.

The partition function

$$Z_N = \sum_{\{n\}} \exp -\beta \left(\mu N + H(n) \right)$$

of a system with fixed number of particles may be described by the expression

$$
Z_N = \sum_{\{n\}} \exp \left\{ \beta \left[\sum_s (\mu - \varepsilon_s) n_s + \nu^2 \frac{1}{2} \sum_{s,s'} W_{ss'} n_s n_{s'} \right. \right.
$$

$$
\left. \left. - \nu^2 \frac{1}{2} \sum_s U_s n_s^2 \right] \right\}, \tag{5.61}
$$

where $\sum_{\{n\}}$ implies summation over all probable distributions $\{n_s\}$, $\beta \equiv 1/kT$, and T is the absolute temperature. The summation in (5.2) may be formally carried out by introducing auxiliary field variables and making use of the known properties of the Gauss integrals [8, 10, 11, 13, 14], i.e.,

$$
\exp \left\{ \frac{\nu^2}{2\theta} \sum_{s,s'} \omega_{ss'} n_s n_{s'} \right\} = \int_{-\infty}^{\infty} D\varphi \exp \left\{ \nu \sum_s n_s \varphi_s \right.
$$

$$
\left. - \frac{\theta}{2} \sum_{s,s'} \omega_{ss'}^{-1} \varphi_s \varphi_{s'} \right\} \tag{5.62}
$$

with

$$
D\varphi = \frac{\prod_s d\varphi_s}{\sqrt{\det (2\pi \beta \omega_{ss'})}}.
$$

Here $\omega_{ss'}^{-1}$ is the inverse matrix that satisfies the equation $\omega_{ss''}^{-1} \omega_{s''s'} = \delta_{ss'}$ and $\nu^2 = \pm 1$ depending on the sign of the interaction or the potential energy. Substituting this presentation of the Gaussian integral in the partition function, we reduce the calculation of the partition function to the quantum field theory procedure with respect to the auxiliary field introduced. This approach provides a derivation of the partition function even for the case of spatially inhomogeneous particle distributions. However, we obtain the partition function in a somewhat different way. We introduce the generating functional that, within the context of the partition function, is given by

$$
\Pi(\mu) = \frac{Z_N(\mu, \varepsilon_s)}{Z_N(\varepsilon_s)},
$$

where

$$Z_N(\mu, \varepsilon_s) = \int D\varphi \exp\left\{-\frac{1}{2\beta}\sum_{s,s'}W_{ss'}^{-1}\varphi_s\varphi_{s'}\right\}Q(\mu, \varepsilon, \varphi), \quad (5.63)$$

and

$$Q(\mu, \varepsilon, \varphi) = \sum_{\{n\}}\exp\left\{\beta\left[\sum_s(\beta\mu - \beta\varepsilon_s + \nu\varphi_s)n_s \right.\right.$$
$$\left.\left. -\nu^2\frac{1}{2}\sum_s U_s n_s^2\right]\right\} \quad (5.64)$$

is the partition function for fixed number of particles, and

$$Z_N(\varepsilon_s) = \int D\psi \exp\left\{-\frac{1}{2\beta}\sum_{s,s'}W_{ss'}^{-1}\psi_s\psi_{s'}\right\}Q(\varepsilon, \varphi) \quad (5.65)$$

is the partition function without chemical potential μ, where

$$Q(\varepsilon, \varphi) = \sum_{\{n\}}\exp\left\{\beta\left[\sum_s(-\beta\varepsilon_s + \nu\psi_s)n_s - \nu^2\frac{1}{2}\sum_s U_s n_s^2\right]\right\}.$$
$$(5.66)$$

The physical meaning of the generating function is obvious from the definition, since it is described by the ratio of the partition function with fixed number of particles to the partition function of the grand canonical ensemble. Now we substitute

$$\varphi_s = \psi_s + \varphi_s^0$$

and

$$n_s = m_s + \nu^2\frac{\varphi_s^0 + \beta\mu}{\beta U_s}$$

in the denominator of the generating functional, then the latter reduces to

$$\Pi(\mu) = \exp\left(\sum_s\beta(\mu - 2\varepsilon_s)\frac{\varphi_s^0 + \nu^2\beta\mu}{2U_s}\right) \quad (5.67)$$

under the condition that φ_s^0 satisfies the equation

$$\varphi_s^0 - U_s \sum_{s'} W ss'^{-1} \varphi_{s'}^0 + \nu^2 \beta \mu = 0. \tag{5.68}$$

The shift of the field variable and the equation obtained select the states that contribute the most in the partition function of the system. Having found the solution of this equation, we can write the generating functional and thus obtain all relevant thermodynamic characteristics of the system. The chemical potential is determined by the equation

$$N = \langle N \rangle = \frac{1}{\beta} \frac{\partial \ln Z_N(\mu, \varepsilon_s)}{\partial \mu} = \frac{1}{\beta} \frac{\partial \ln \Pi}{\partial \mu} \tag{5.69}$$

and the compressibility of the system is given by

$$\varkappa = \frac{N^2}{\beta V} \left(\frac{\partial V}{\partial P} \right) = \frac{1}{\beta} \frac{\partial \langle N \rangle}{\partial \mu} = \frac{1}{\beta} \frac{\partial^2 \ln \Pi}{\partial \mu^2}, \tag{5.70}$$

where P is the pressure, and V is the volume of the system. This becomes possible after the first step of the change of variables, which reduces the partition function

$$Z_N(\mu, \varepsilon_s) = \Pi(\mu) Z_N(\varepsilon_s)$$

to the product of two factors. We employ the presentation $Z_N(\varepsilon_s)$, consider again a new shift of variables

$$\psi_s = \sigma_s + \nu \xi_s, \quad m_s = k_s + \nu^2 \frac{\nu \xi_s + \varepsilon_s \beta}{\beta U_s},$$

and require that ξ_s should be a solution to the equation

$$\xi_s - U_s \sum_{s'} W_{ss'}^{-1} \xi_{s'} - \beta \varepsilon_s = 0. \tag{5.71}$$

Then the partition function of the grand canonical distribution of the system reduces to the form

$$Z_N(\varepsilon_s) = \Xi(\varepsilon_s) Z_N(0),$$

where

$$\Xi(\varepsilon_s) = \exp\left(\sum_s (\beta\varepsilon_s - \xi_s)\frac{\varepsilon_s}{2U_s}\right), \tag{5.72}$$

and

$$Z_N(0) = \int D\sigma \exp\left\{-\frac{1}{2\beta}\sum_{s,s'} W_{ss'}^{-1}\sigma_s\sigma_{s'}\right\}$$

$$\times \sum_{k_s} \exp\left\{i\sigma_s k_s + \frac{1}{2}\nu^2\beta U_s k_s^2\right\} \tag{5.73}$$

describes the configuration partition function of the system with the external field disregarded. It is not difficult to see that the last term in the formula is the Riemann Θ function doubly periodic with respect to the variables $i\sigma_s/2\pi$ and $(i\nu^2\beta U_s)/2\pi$. This function is tabulated for the whole range of change of arguments. Now the configuration partition function

$$Z_N(0) = \int D\sigma \exp\left\{-\frac{1}{2\beta}\sum_{s,s'} W_{ss'}^{-1}\sigma_s\sigma_{s'}\right\}\Theta\left(-\frac{i\sigma_s}{2\pi}, \frac{i\nu^2\beta U_s}{2\pi}\right) \tag{5.74}$$

is written in terms of the well-known Θ functions. Thus, the partition function of the system is completely determined for the statistical description of such systems. In this sense we may say that the procedure provides an exact statistical solution for a system of interacting particles described by the present Hamiltonian. For a system with unfixed number of particles, formulas thus obtained yield the thermodynamic characteristics of the system, i.e., the correlation function and the susceptibility. We have

$$g_{ss'} = \frac{\partial^2 \ln \Xi(\varepsilon_s}{\partial\varepsilon_s\partial\varepsilon_s'}, \quad \varkappa = \frac{1}{N\beta}\sum_{s,s'} g_{ss'}. \tag{5.75}$$

Thus, we have a complete statistical description of the system by the known Θ functions and solutions of the relevant equations for

φ_s and ξ_s. Solutions of the equations for these variables select the configuration states that dominates the most in the partition function. To obtain the final result requires specifying the interaction $W_{ss'}$ and U_s. We rewrite the generating functional in a modified form as

$$\Pi(\mu) = \exp\left(\nu^2 \sum_s (\mu - 2\varepsilon_s)\beta\mu f_s\right), \tag{5.76}$$

where

$$f_s = \frac{1}{U_s}\left(1 - \sum_s{}' G_{ss'}\right)$$

is written in terms of the Green function of the equation

$$G_{ss'} - U_s \sum_s{}'' G_{ss''} = \delta_{ss'}. \tag{5.77}$$

This equation reduces the equation for the chemical potential to

$$N = \sum_s (\mu - \varepsilon_s) f_s. \tag{5.78}$$

The compressibility of the system is described by the expression

$$\varkappa = \frac{N}{\beta V}\left(\frac{\partial V}{\partial P}\right) = \frac{\nu^2}{\beta}\sum_s f_s. \tag{5.79}$$

So, in order to find the thermodynamic characteristics of the system one just has to find the solution for the Green function $G_{ss'}$. The above treatment is based on the initial lattice model and has an exact solution for this case. It is not difficult to extend the approach to the continuum case. To do this, we cannot avoid calculating the inverse operator $W_{ss'}^{-1}$. If the interaction energy is given by $W_{ss'} = W(R_s - R'_s)$ (the values of the subscript s form a continuous set within the volume V of the system), then the inverse matrix should be interpreted in the operator sense, i.e., $W_{ss'}^{-1} = \delta(R - R')L_R$, where L_R is the operator for which the interaction potential is given by the

Green function. For screening Coulomb or Newtonian interaction we have

$$W_{ss'} = \frac{q^2}{R_s - R'_s} \exp\left[-\lambda(R_s - R'_s)\right],$$

where λ is the screening length and q is the interaction constant

$$L_R = -\frac{1}{4\pi q^2}(\Delta_R - \lambda^2)$$

with Δ_R being the Laplace operator. This representation provides a possibility to obtain the equation of state for the continuum case, when interaction at different sites is Newton attraction or Coulomb repulsion and interaction at the same site is constant, $U = \mathrm{const}$. The passage to the continuum limiting case is carried out by replacing

$$\sum_s = \frac{N}{V}\int dV,$$

where $N/V = \rho$ is the density, and rewriting the equations in the differential form. The equation for the Green function in the continuous case is given by

$$\rho G(R) + \frac{U}{4\pi q^2}(\Delta_R - \lambda^2)G(r) = \delta(R). \qquad (5.80)$$

In this case λ^2 is greater than $(4\pi q^2)/U\rho$ and the solution of this equation is given by

$$G(R) = -\frac{q^2}{UR}\exp(-\lambda_0 R),$$

where $\lambda_0^2 = \lambda^2 - 4\pi q^2/U\rho$. If the reversed inequality holds, then

$$G(R) = -\frac{q^2}{UR}\cos(\lambda_0 R).$$

Substituting the Green function in the definition of the generating functional and making use of the continuum analogue yields

$$\int f(R)d^3R = -\frac{N}{\beta V}\left(\frac{\partial V}{\partial P}\right).$$

Thus we find that for the first case

$$\left(\frac{\partial \rho}{\partial P}\right)_T = \frac{\nu^2}{U}\left(1 + \frac{\rho^2}{\rho_0^2}\right),$$

where

$$\rho_0^2 = \frac{N^2 \lambda^5 U}{4\pi q^2}.$$

The pressure increase over the lattice gas pressure $P_0 = kT\rho$ is described by a simple expression, i.e.,

$$\delta P = P - P_0 = \frac{U}{\nu^2}\rho_0 \arctan\left(\frac{\rho}{\rho_0}\right),$$

that for ρ/ρ_0 reduces to

$$\delta P = P - P_0 = \frac{U}{\nu^2}\rho\left(1 - \frac{1}{3}\frac{\rho^3}{\rho_0^3}\right).$$

If the reversed inequality is satisfied, then the pressure increase may be obtained in the integral form. If the interaction at different lattice sites is constant, $W_{ss'}^{-1} = W = \text{const}$, we have $\delta P = \nu^2(U - W)\rho$ which provides a complete physical picture of the situation under consideration. The equation of state is not of van der Waals form; nevertheless, it allows a phase transition for $kT_c = -U/\lambda^2$ in the case $\nu^2 = -1$. The phase transition in the system with constant interaction energy occurs for

$$kT_c = \nu^2(U - W).$$

To conclude, we reduce the partition function of the three-dimensional lattice system with long-range and short-range interactions at the sites to the form that can be calculated by conventional numerical methods. For instance, the particle distribution function may be obtained as a result of such calculations. However, the conventional thermodynamic characteristics (pressure, compressibility, etc.) and the equation of state that do not imply knowledge of the distribution function, may be obtained even without such calculations, due to the factorization of the partition function.

We emphasize that our choice of the interaction and dimension three is dictated by the fact that the inverse operator has the form that is essential to our approach in just three dimensions. However, since other potentials have the same inverse operator with $\lambda = 0$, respectively, in one, two, and three dimensions, the method may also be applied to the relevant systems with dimensions other than three.

5.5   Gravitating Gas Model

Let us consider a system of particles whose interaction consists of gravitational attraction and hard spheres repulsion. For the Newtonian attraction, the inverse operator is known to be given by

$$ W_{rr'}^{-1} = -\frac{1}{4\pi Gm^2}\,\Delta_{r'}\delta_{rr'}, $$

where G is the gravitation constant, m is the particle mass, and Δ_r is the d'Alembert operator. We employ the results of the hard spheres model and thus obtain an expression for the action, i.e.,

$$ S = \int_{V_0} dV \left\{ \frac{1}{4r_m}(\nabla\varphi)^2 - \xi A e^{\varphi} \right\} + \xi A V_0 + (N+1)\ln\xi, \quad (5.81) $$

where $r_m \equiv 2\pi Gm^2\beta$, and integration is carried out over the whole space except for the volume occupied by particles. An expression, analogous to (5.81), was obtained in [8]. The authors, however, did not fix the number of particles and disregarded particle repulsion. The result of [8] may be supplemented with the solutions that allow for inhomogeneous particle distributions.

We introduce a new dimensionless quantity $r = R/r_m$ and denote $\gamma^2 \equiv \xi A r_m^3$. Then the action (in spherical coordinates) may be written using of a new variable $\sigma = \exp(\varphi/2)$, i.e.,

$$ S = 4\pi \int_{r_0}^{\infty} \left\{ \left(\frac{1}{\sigma}\frac{d\sigma}{dr}\right)^2 - \gamma^2\sigma2 \right\} r^2 dr + \xi A V_0 + (N+1)\ln\xi. $$

$$ (5.82) $$

If the cluster surface contribution in the action (5.82) is negligible, then the term $(2/r)(d\sigma/dr)$ may be omitted [160]. Hence the saddle-point equation reduces to

$$\frac{d^2\sigma}{dr^2} - \frac{1}{\sigma}\left(\frac{d\sigma}{dr}\right)^2 + \gamma^2\sigma^3 = 0. \tag{5.83}$$

The first integral of this equation is given by

$$\left(\frac{1}{\sigma}\frac{d\sigma}{dr}\right)^2 + \gamma^2\sigma^2 = \Delta^2, \tag{5.84}$$

where Δ^2 is an unknown integration constant. It should be noted that the first integral of Eq. (5.83) is similar to that of the nonlinear Schrödinger equation. The solution of Eq. (5.83) with the first integral (5.84) is given by

$$\sigma = \frac{\Delta}{\gamma}\frac{1}{\cosh \Delta r}. \tag{5.85}$$

Thus, introducing the ansatz $\varphi = \ln \sigma^2$ enabled us to find the solution of the nonlinear equation

$$\frac{1}{2}\frac{d^2\varphi}{dr^2} + \gamma^2 e^\varphi = 0,$$

that satisfies the extremum condition of the functional

$$S = \int r^2 dr \left\{\frac{1}{4}(\nabla\varphi)^2 - \gamma^2 e^\varphi\right\}.$$

When considering the one-dimensional solution, we have assumed that the variation range of φ or σ is shorter than the dimension of the soliton formation described by the solution (5.85). In our interpretation, any soliton solution corresponds to a spatially inhomogeneous particle distribution, i.e., a finite-size cluster. It depends on the interaction parameters, chemical potential, and temperature, which of these solutions is realized. In the model under consideration, the soliton solution is associated with the case when, by virtue of gravitational attraction, particles are concentrated within a volume limited by their sizes. This corresponds to the solution (5.85) with the asymptotics $\sigma^2 = 1$, $\varphi = 0$ for $r = d$, where d is the cluster size,

$\sigma \to 0$, $\varphi \to -\infty$ as $r \to \infty$. Physically, this solution describes the occurrence of particles in the inhomogeneous formation of the size d and the absence of particles at infinity, since in this case the spatial distribution function is $f(r) = \xi A e^{\varphi}$. Within the context of (5.85), action (5.82) may be rewritten in the form

$$S = 4\pi \int_{r_0}^{d} \left(\Delta^2 - 2\gamma^2\sigma^2\right) r^2 dr + \gamma^2 \frac{V_0}{r_m^3} + (N+1)\ln \frac{\gamma^2}{Ar_m^3}. \quad (5.86)$$

Now we perform integration

$$2\gamma^2 \int_{r_0}^{d} \sigma^2 r^2 dr = 2\Delta^2 \int_{r_0}^{d} \frac{r^2 dr}{\cosh^2 \Delta r}$$

and obtain

$$2\gamma^2 \int_{r_0}^{d} \sigma^2 r^2 dr = f(\Delta) - g(\Delta), \quad (5.87)$$

where

$$f(\Delta) = \frac{2}{\Delta} \left\{ \Delta^2 d^2 \tanh \Delta d - \Delta^2 r_0^2 \tanh \Delta r_0 \right\} \quad (5.88)$$

and

$$g(\Delta) = \frac{2}{\Delta} \left\{ 2 \sum_{k=1}^{\infty} \frac{2^{2k} \left(2^{2k} - 1\right)}{(2k+1)(2k)!} \left[B_{2k} \left(\Delta d\right)^{2k+1} - B_{2k} \left(\Delta r_0\right)^{2k+1} \right] \right\}. \quad (5.89)$$

Here B_{2k} are the Bernoulli numbers. We expand the result in power series of $\Delta d \ll 1$. Then the action (5.86) reduces to

$$S = -\frac{\Delta^2}{r_m^3} (V - V_0) + \gamma^2 \frac{V_0}{r_m^3} + (N+1)\ln \frac{\gamma^2}{Ar_m^3}. \quad (5.90)$$

As $\sigma^2 = 1$ for $r = d$, we have

$$\gamma^2 \simeq \Delta^2 \left(1 - \Delta^2 d^2\right).$$

Substituting the latter expression in (5.90), we find the action to be given by

$$S = \Delta^2 \frac{V_0}{r_m^3} - \Delta^2 \frac{V - V_0}{r_m^3} + (N+1)\ln \frac{\Delta^2}{Ar_m^3}$$

$$+ (N+1)\ln\left(1 - \Delta^2 d^2\right) - \frac{V_0}{r_m^3}\Delta^4 d^2. \tag{5.91}$$

In the next step, we find the extremum of (5.91) with respect to Δ by the iteration method. Thus, we obtain

$$\tilde{\Delta}^2 = r_m^3 \frac{N+1}{V - 2V_0}.$$

We substitute the latter expression in (5.91) to obtain the action of the form

$$S = -(N+1) + (N+1)\ln \frac{N+1}{AV\left[1 - 2V_0/V + \tilde{\Delta}^2 d^2\left(1 - 2V_0/V\right)\right]}. \tag{5.92}$$

Now we have to find d. In what follows we shall see that its value may be found by minimizing the action. Here, however, we estimate it in light of physical reasoning. Let us consider the condition for particle confinement within a cluster. For $r = 2r_0$, we have $\varphi = \varphi_0$ and hence

$$d = 2r_0 + \frac{\varphi_0}{2\tilde{\Delta}}.$$

If particles are not confined within the cluster, $d \simeq 2r_0$ and thus

$$\tilde{\Delta}^2 d^2 \left(1 - \frac{2V_0}{V}\right) = \frac{(N+1)r_m^3 4r_0}{V} = \frac{V_0}{V}\frac{6Gm^2}{r_0}\beta.$$

Therefore, within the context of (5.92), the partition function of the system is given by

$$Z_N = Z_N^0 \left[1 - \frac{2V_0}{V} + \frac{V_0}{V}\frac{6Gm^2}{r_0 kT}\right]^{N+1}. \tag{5.93}$$

It should be noted that the gravitating-gas partition function thus obtained exactly reproduces the known expression [53, 54] that was derived with the use of virial coefficients. The approach proposed here is more general because it makes it possible to estimate the partition function in the presence of clusters of arbitrary sizes $R = dr_m$, i.e.,

$$Z_N = Z_N^0 \left[1 - \frac{2V_0}{V} + \frac{R^2 r_m (N+1)}{V} \right]^{N+1}, \qquad (5.94)$$

where Z_N^0 is the partition function of the ideal Boltzmann gas.

5.6   Coulomb-like Systems

One of the ways to describe spatially inhomogeneous distributions in a system of interacting particles is to use the new unconventional method proposed in [3, 121] that employs the Hubbard–Stratonovich representation of the partition function. Now this method is extended and applied to a system with Coulomb-like interaction to find the solution for the particle distribution. It is important that this solution has no divergences for the thermodynamic limiting cases. The purpose of this section is to apply the quantum-field-theory approach to the statistical description of a Coulomb-like system and to calculate the thermodynamic characteristics for both homogeneous and inhomogeneous distributions of interacting particles. The condensed spatially periodic structures are studied as well. We consider electrically neutral systems and use the pure Coulomb interaction potential or an effective screened potential, provided an appropriate model can be introduced. We use the saddle-point approximation for the conservation of the number of particles that yields a nonlinear equation for the new field variable. In the three-dimensional case, this equation reduces to the sine-Gordon equation whose solution determines the state associated with the most contribution in the partition function. The method makes it possible to describe the conditions of Wigner crystal formation in a system of dust particles in a plasma. Various possibilities may be associated with different parameters corresponding to the interaction potential. Nevertheless, the results for simple and basic cases are most important for understanding the behavior of dusty plasmas in complex situations. The necessary condition for the crystal formation in a system of dust particles is derived analytically for the three-dimensional case. For the one- and two-dimensional cases, exact solutions for various spatial distributions of charged particles are found.

In the continuum case the free energy of a system of particles with Coulomb-like interaction may be written in the form

$$\beta F = \int dV \left\{ \frac{1}{8\pi Q^2 \beta} (\nabla \varphi)^2 + \kappa^2 \varphi^2 \right.$$

$$\left. + \delta \sum_p \ln \left(1 - \delta \xi e^{-\beta \varepsilon_p} \cos \varphi \right) \right\}.$$

As has been shown in [53, 54] for the classical statistics, we have ξ less than one and

$$\ln \left(1 - \delta \xi e^{-\beta \varepsilon_p} \cos \varphi \right) = \delta \xi e^{-\beta \varepsilon_p} \cos \varphi.$$

The effective free energy for the Boltzmann statistics can be rewritten in the form [121]

$$\beta F = \int dV \left\{ \frac{1}{2r_e} \left((\nabla \varphi)^2 + \kappa^2 \varphi^2 \right) - \xi A \cos \varphi \right\} + (N + 1) \ln \xi.$$

where $r_e = 4\pi Q^2 \beta$ and

$$A \equiv \lambda^{-3} = \left(\frac{2\pi m}{\beta h^2} \right)^{3/2}$$

after integration over all momenta.

One-dimensional case

Now we consider a one-dimensional system with linear particle density. Charges distributed along a macromolecule may be regarded as an example of such a system. We consider a cylindrical molecule of length L and radius r shorter than the length. Let the Coulomb charges lie on the cylinderical axis. In this case, the problem can be solved exactly [121]. The free energy of a system of charged particles in the one-dimensional case may be presented as

$$\beta F = \frac{V}{L} \int_0^L dz \left\{ \frac{1}{r_e} \left(\frac{d\varphi}{dz} \right)^2 - \xi A \cos \varphi \right\} + (N+1) \ln \xi.$$

The saddle-point equation reduces to the sine-Gordon equation

$$\frac{1}{r_e} \left(\frac{d^2 \varphi}{dz^2} \right) + \xi A \sin \varphi = 0$$

with the first integral given by

$$\frac{1}{r_e} \left(\frac{d\varphi}{dz} \right)^2 + \xi A \cos \varphi = C.$$

The relevant exact solution is given with finite period, i.e.,

$$l = \frac{1}{\sqrt{r_e}} \int \frac{d\varphi}{\sqrt{C - \xi A \cos \varphi}} = \frac{K(p)}{\sqrt{2 r_e (C + \xi A)}},$$

where $K(p)$ is the full elliptic integral of the first kind with the unknown argument

$$p = \sqrt{\frac{2\xi A}{C + \xi A}}.$$

This solution depends on the integration constant [60]. Substituting the solution into the free energy yields

$$\beta F = 2\xi AV \left\{ \frac{2E(p)}{p^2 K(p)} - \frac{1}{p^2} + 1 \right\} - \xi AV + (N+1) \ln \xi. \qquad (5.95)$$

Here $E(p)$ is the full elliptic integral of the second kind with the same argument. The free energy extremum is attained for $p = 1$ or $C = \xi A$ which corresponds to the soliton solution given by

$$\varphi = 4 \arctan \exp(z \sqrt{r_e \xi A}).$$

The latter solution determines the state associated with the most contribution in the partition function. Thus, the free energy takes the form

$$\beta F = \beta F_B + 8 \left[(N+1) \frac{L}{r_e} \right]^{1/2}.$$

In this case the free energy of interacting particles grows with the increase of the number of particles and the size of the system. The probable periodic structure may be motivated by the spatial boundary conditions. The period of the structure

$$l = L \left(\frac{L}{(N+1)r_e} \right)^{1/2} \tag{5.96}$$

increases with the decrease of the number of particles. Such a system is homogeneous on the macroscopic scale, however, the particle distribution can be periodic.

Two-dimensional case

As was shown in [51], the Debye–Huckel theory for the interfacial geometry predicts the Coulomb-like systems for charged colloids or charged polymers at monolayers, solid substrates, and interfaces. The phase diagram of two-dimensional electron liquids was described in [61–65, 71, 76]. In the case of a two-dimensional system we can obtain an exact solution for the homogeneous distribution of particles. We consider the Coulomb-like potential

$$\Omega_{i,j} = \frac{Q^2}{r_e} \ln k r_{i,j},$$

where $r_{i,i}$ is the distance between particles, and r_e is the average distance between particles. For charged colloidal particles we have $Q^2 = l_b \rho^2$, where $l_b = (e^2)/(4\pi \epsilon k T)$ is the Bjerrum length that defines the length along which two unit charges interact with thermal energy, and ρ is the charge density. This model potential has the form similar to the interaction between two homogeneously charged lines in the three-dimensional space. Thus, the motion of actual charges on the two-dimensional plane is similar to the motion of parallel lines oriented perpendicularly to the plane of the two-dimensional system.

In the continuous limiting case the partition function for the system under consideration may be written in the standard form

$$Z_N = \int \exp\left[-\beta H(r,p)\right] d^N \mathbf{r} d^N \mathbf{p},$$

where Hamiltonian of the system is given by

$$H(r,p) = \sum_i \frac{p_i^2}{2m} + \frac{1}{2} \sum_{i,j} \omega_{ij}$$

with the potential energy being given by

$$\omega_{ij} = \frac{Q^2}{\langle r \rangle} \ln \varkappa r_{ij}.$$

In the two-dimensional case with homogeneous distribution of particles we can obtain the exact description by the equation of state obtained, i.e.,

$$P = kT \frac{\partial \ln Z_n}{\partial S},$$

where S is the square of a circle of radius R. Having introduced the dimensionless variable $r_i' = r_i/S^{1/2}$ we can rewrite the partition function in the form

$$Z_N = S^N \int \exp\left[-\beta H(r,p)\right] d^N \mathbf{r}' d^N \mathbf{p}', r_i' = \frac{r_i}{S^{1/2}}.$$

The derivative of the partition function in this case may be presented as

$$\frac{\partial Z_N}{\partial S} = \frac{NZ}{S} - \frac{S^N}{kT} \int \exp\left[-\beta H(r,p)\right] \frac{1}{2} \sum_{i,j} \frac{r_{ij}}{2S} \frac{\partial \omega_{ij}}{\partial r_{ij}} d^N \mathbf{r}' d^N \mathbf{p}'.$$

This result may be presented in the dimensional form within the context of the explicit form of the potential

$$\frac{\partial Z_N}{\partial S} = \frac{NZ}{S} - \frac{N(N-1)Z}{4SkT}.$$

Substituting this result into the equation of state generates the exact solution for the two-dimensional Coulomb-like system, i.e.,

$$PS = NkT\left\{1 + \frac{(N-1)Q^2}{4kT\langle r\rangle}\right\}.$$

If we take into account that $\langle r\rangle n^{1/2}$, where n is the concentration of particles, then it is possible to show that, for large concentrations, the system of interacting particles is unstable and the inhomogeneous particle distribution can arise. Such inhomogeneous distributions are caused by the long-range nature of the Coulomb interaction. In the case of strong interaction, the Coulomb-like system is unstable as a whole, so the minimum value of the free energy is attained for inhomogeneous distributions of particles. Now we employ the proposed approach in order to find the states associated with the Wigner crystal. In the two-dimensional case, we may write the effective free energy in a simple form given by

$$\beta F = \frac{V}{S}\int dxdy\left\{\frac{1}{r_e}\left((\nabla\varphi)^2 + \varkappa^2\varphi^2\right) - \xi A\cos\varphi\right\} + (N+1)\ln\xi,$$

where h is the thickness of the two-dimensional layer. In the general case, the equation for the saddle-point states is given by

$$\frac{1}{r_e}\left\{\Delta_2\varphi - \varkappa^2\varphi\right\} + \xi A\sin\varphi = 0, \quad \frac{V}{S}\int dxdy\xi A\cos\varphi = N+1,$$

where Δ_2 is the two-dimensional Laplace operator. The chemical activity may be obtained from the normalization condition

$$\frac{1}{r_e}\left[(\nabla\varphi)^2 - \varkappa^2\varphi^2\right] + \xi A\cos\varphi = E,$$

where E is an unknown integration constant that should be found from the existence of the solution. Although this equation cannot be solved in the general case, it provides a tool to study many interacting Coulomb-like systems under various external conditions. In terms of the density function $\rho(r)$ the first integral may also be written as

$$\beta F = \frac{V}{S} \int dx dy \left\{ E + 2 \frac{\varkappa^2}{r_e} \arccos^2 \left(\frac{\rho}{\xi A} \right) - 2\rho \right\} + (N+1) \ln \xi$$

and the effective free energy in terms of the first integral reduces to the form

$$\beta F \simeq \beta F_B + (N+1) \left\{ \frac{\pi^2 \varkappa^2}{4nr_e} - 1 \right\}.$$

Let us consider the case of a periodic distribution of particles in the system. Here, we have to assume that $\rho = \rho_0(1 + \cos(kr))$ where r is the surface variable and k is the reciprocal lattice vector of the periodic distribution. The first integral should be constant for each point of the space including the space point where $\rho = 0$. In this case, the first integral may be determined as

$$E = -\frac{\pi^2 k^2}{4r_e}.$$

Substituting this relation into the free energy, we can present the free energy as

$$\beta F = V \left\{ -\frac{\pi^2 k^2}{4r_e} + 2 \frac{\varkappa^2}{r_e} \arccos^2 \left(\frac{\rho_0}{\xi A} \right) \right\} + (N+1) \ln \xi,$$

where the value of the integral over the coordinate space is estimated in terms of the averaged concentration and $V = hS$ is the volume of the system. From the condition of minimum effective free energy we conclude that the solution exists for $\xi A = \rho_0$ and the effective free energy takes the form

$$\beta F = \beta F_B + (N+1) \left\{ \frac{\pi^2 k^2}{4nr_e} - 1 \right\}.$$

This quantity can be smaller than the free energy of gas for

$$\frac{\pi^2 k^2}{4nr_e} < 1.$$

An example of such a system is given by the electrons on the helium surface where we have repulsive electrostatic interaction

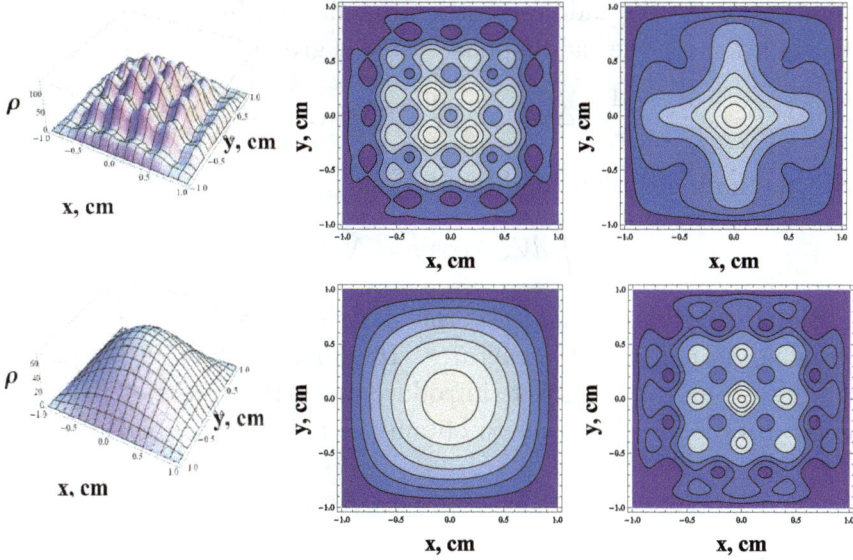

Fig. 5.1.   Structure of electron distribution on the helium surface [62].

between electrons in the form

$$V_e(r) = \frac{e^2}{4\pi\varepsilon_0 r} - \frac{4\Lambda_s}{\sqrt{r^2 + (2d)^2}}, \quad \Lambda_s = \frac{\varepsilon_s - \varepsilon_{He}}{16\pi\varepsilon_0(\varepsilon_s + \varepsilon_{He})}e^2$$

and the attractive capillary interaction is given by

$$V_c(r) = a(E^2)K_0\big(\lambda(\mathbf{r} - \mathbf{r}')\big).$$

If we carry out computer simulation for the determination of the behavior of a system of electrons on the helium surface (please see Fig. 5.1), we may obtain a real period of the structure. This result was published in [62].

Three-dimensional case

The three-dimensional structure of electro-rheological solids was described in [62, 63]. In this section, we describe a system of dust particles in a weakly ionized plasma in the stationary case. In this case, we again use a method that makes it possible to determine the states with the most contributions to the partition function.

Namely, we use the saddle-point approximation. In the general case, the equation for the saddle-point states is given by

$$\frac{1}{r_e} \left\{ \Delta \varphi - \varkappa^2 \varphi \right\} + \xi A \sin \varphi = 0,$$

with the normalization condition

$$\int dV \xi A \cos \varphi = N + 1.$$

In the three-dimensional case the description takes into account probable formation of structures. It should be noted that this condition makes it possible to introduce the concentration of particles as $\rho = \xi A \cos \varphi(r)$ and thus to obtain the first integral given by

$$\frac{1}{r_e} \left\{ (\nabla \varphi)^2 + \varkappa^2 \varphi^2 \right\} + \xi A \cos \varphi = E, \tag{5.97}$$

where E is an unknown integration constant. Similarly to the two-dimensional case, this equation cannot be solved explicitly in the general case. We start from a homogeneous distribution of interacting particles. In this case we have to derive the condition for the existence of the solution $\varphi = \varphi_0 = \text{const}$ from the equation for the field variable

$$\frac{k^2}{r_e} + \xi A \cos \varphi_0 = E$$

and to find the chemical activity from the normalization condition

$$\xi A \cos \varphi_0 V = N + 1.$$

Within the context of the first integral and the equation for the chemical activity, the free energy may be written as

$$\beta F = \beta F_B + (N + 1)^{1/3} - \ln \cos \varphi_0,$$

where we have introduced the concentration of particles $n = (N + 1)/V$. Obviously, the second term in the free energy is always positive because $\cos \phi_0$ and thus the free energy of a homogeneous system of interacting particles is greater than the free energy of the Boltzmann gas. In the general case, particle distributions in Coulomb-like systems are inhomogeneous, the long-range nature of the Coulomb interaction being the reason. Similarly to the two-dimensional Coulomb-like system in the case of strong interaction, the system becomes unstable as a whole, so the minimum value of the free energy is attained in the case of inhomogeneous distribution of particles. Now we employ the proposed approach to show how to find the states associated with the Wigner crystal. Bearing in mind that the density function $\rho = \xi A \cos \phi$ is only positive and assuming that the state of interest does exist, we take the periodic distribution function in the form

$$\rho(\mathbf{r}) = \xi A \cos \varphi = \xi A \left\{ 1 + \cos(k_x x) + \cos(k_y y) + \cos(k_z z) \right\},$$
$$k = 2\pi n^{1/3}$$

that corresponds to the cubic lattice with the wave vector with components k_x, k_y, k_z. If we assume that one charged particle is present at each lattice site and that the lattice is isotropic, $k_x = k_y = k_z = n^{1/3}$ is the particle density, then the normalization condition yields or $n_s = (N+1)/AV$. From the first integral for the field variable, one can conclude that $E^2 = (\pi^2 \lambda^2)/(4r_e)$.

We substitute this relation into the free energy written in terms of the first integral, i.e.,

$$\beta F = \beta F_B + (N+1) \left\{ \frac{\pi^2 \varkappa^2}{4nr_e} - 1 \right\}.$$

Introducing the coupling parameter $\Gamma_e \equiv r_e n^{1/3}$ (that is the ratio of the Coulomb to kinetic energy) provides a relation for the critical value of the coupling parameter, i.e.,

$$\Gamma_e \geq 4\pi^2 \varkappa^2 n^{2/3} \equiv (2\pi \varkappa L)^2.$$

Namely, such structures are observed in dusty plasmas. In terms of the structure lattice parameter used in [34], $l = kL$ (the interparticle distance normalized by the effective screening length), the relation obtained is given by

$$\Gamma_e \geq (2\pi l)^2, \quad l \equiv \varkappa L,$$

where L is the lattice period. With this condition being satisfied, we may expect a crystal structure to be formed. This relation gives the value of the same order as the result of the numerical simulation [34]. We cannot solve the problem of crystal structure formation in dusty plasmas exactly. Nevertheless we can analytically predict the conditions for such formation. The problem is that we do not know the three-dimensional solution of the sine-Gordon equation that determines the field variable. Moreover, this approach provides a description of spatially periodic distributions. The partition function has no singularities for any values of the Coulomb-like field. It is shown that the minimum of the free energy sometimes does not correspond to a homogeneous particle distribution, but could indicate the formation of a crystal-like structure.

Chapter 6

Inhomogeneous Distribution in Systems of Particles

6.1 Microcanonical Description of Gravitating Systems

Now we consider the microcanonical evolution of a self-gravitating system. In the case of a microcanonical ensemble, the object of fundamental interest is the density of states that may be written in the standard form [53] given by

$$\Omega_{E,N} = \int d\mathbf{q} d\mathbf{p} \delta[H(\mathbf{p}, \mathbf{q}) - E] \delta[N(\mathbf{p}, \mathbf{q}) - N], \qquad (6.1)$$

where $\Omega_{E,N}$ denotes the microcanonical partition function for fixed energy and number of particles in the system. We introduce the Lagrange multiplier $\beta = 1/kT$, i.e., the reciprocal temperature, and $\eta = \beta\mu$, where μ is the chemical potential, and perform the Laplace transformation. Then the microcanonical distribution function may be presented in the form given by

$$\Omega_{E,N} = \oint d\beta \oint d\eta \exp\{\beta E - \eta N\}$$

$$\times \int d\mathbf{q} d\mathbf{p} \exp\{-\beta H(\mathbf{p}, \mathbf{q}) + \eta N(\mathbf{p}, \mathbf{q})\}. \qquad (6.2)$$

The entropy and temperature for the microcanonical ensemble are determined by $S = \ln \Omega_{E,N}$, where the reciprocal temperature may be found from the relation $\beta = \frac{dS}{dE}$. The pressure in the system may be obtained by the formula $P = \frac{1}{\beta}\frac{\partial S}{\partial V}_E$ and thus we come to the equation of state for the system.

A system of interacting particles may be treated in the classical manner similarly to the Ising model with the Hamiltonian [14] that may be presented by the occupation number, i.e.,

$$H(n) = \sum_s \varepsilon_s n_s - \frac{1}{2} \sum_{s,s'} W_{ss'} n_s n_{s'}, \qquad (6.3)$$

where ε_s is the additive part of the energy in the state s that is in most cases equal to the kinetic energy [14], $W_{ss'}$ is the interaction energy for particles in the states s and s'. In this model, macroscopic states of the system are described by a set of occupation numbers n_s. Index s labels an individual particle state and also may correspond to a fixed site on the Ising lattice [54]. Its explicit form in the continuum approximation is irrelevant. The number of particles is fixed and may be found from the relation

$$N(n) = \sum_s n_s.$$

It is clear that the calculation of the partition function remains a rather complicated problem even in the case of the Ising model. The microcanonical partition function of a system of interacting particles may be written as [3]

$$\Omega_{E,N} = \oint d\beta \oint d\eta \exp\{\beta E - \eta N\} \sum_{\{n\}} \exp(-\beta H(n)), \qquad (6.4)$$

or in the explicit form

$$\Omega_{E,N} = \oint d\beta \oint d\eta \exp\{\beta E - \eta N\}$$

$$\cdot \sum_{\{n\}} \exp\left\{-\beta\left[\sum_s \varepsilon_s n_s - \frac{1}{2}\sum_{s,s'} W_{ss'} n_s n_{s'}\right]\right\}, \qquad (6.5)$$

where $\sum_{\{n\}}$ implies summation over all probable distributions $\{n_s\}$. In order to perform the formal summation, we introduce an additional field variable that makes it possible to employ the theory of Gaussian integrals [12, 14], i.e.,

$$\exp\left\{\frac{1}{2\theta}\nu^2 \sum_{s,s'} w_{ss'} n_s n_{s'}\right\}$$

$$= \int_{-\infty}^{\infty} D\varphi \exp\left\{\nu \sum_s n_s \varphi_s - \frac{1}{2\beta} \sum_{s,s'} w_{ss'}^{-1} \varphi_s \varphi_{s'}\right\}, \qquad (6.6)$$

where $D\varphi = \dfrac{\Pi_s d\varphi_s}{\sqrt{\det 2\pi\beta w_{ss'}}}$ and $w_{ss'}^{-1}$ is the inverse of the interaction matrix. The latter satisfies the condition $w_{ss''}^{-1} w_{s''s'} = \delta_{ss'}$ with $\nu^2 = \pm 1$ depending on the character of the interaction energy. If instead of the chemical potential we introduce another important variable, the chemical activity $\xi \equiv e^{\beta\mu} = e^{\eta}$, and using the obvious relation $d\eta = \xi^{-1} d\xi$, then the microcanonical partition function of a system of interacting particles may be rewritten as

$$\Omega_{E,N} = \oint d\beta \oint d\xi \int_{-\infty}^{\infty} D\varphi \exp\left\{\beta E - N\ln\xi - \frac{1}{2\beta} \sum_{s,s'} (W_{ss'}^{-1} \varphi_s \varphi_{s'})\right\}$$

$$\cdot \prod_s \{\xi \exp(-\beta\varepsilon_s + \varphi_s)\}^{n_s}. \qquad (6.7)$$

This microcanonical partition function may be used in the calculation of any thermodynamic property of the system with fixed total number of particles and energy. Such presentation makes it possible to perform summation over the occupation numbers n_s [3], then the microcanonical partition function finally reduces to

$$\Omega_{E,N} = \oint d\beta \oint d\xi \int_{-\infty}^{\infty} D\varphi \exp[\beta S_{\text{eff}}(\varphi, \xi)]. \qquad (6.8)$$

Here, we have introduced the effective entropy

$$S_{\text{eff}}(\varphi, \xi) = \frac{1}{2\beta} \sum_{s,s'} (W_{ss'}^{-1} \varphi_s \varphi_{s'}) - \delta \sum_s \ln\left(1 + \xi e^{\varphi - \beta\varepsilon_s}\right)$$

$$+ (N+1)\ln\xi - \beta E, \qquad (6.9)$$

where $\xi \equiv e^{\beta\mu}$ is the absolute chemical activity of the chemical potential μ. In a proposed calculation, we use the Fermi statistics because two classical particles cannot occupy the same spatial site. After that, we may present the partition function in the usual form $\Omega_{E,N} = \exp S$, where S is the entropy of the system. With this purpose in view, we employ during the microcanonical partition function determination, the efficient methods developed in quantum field theory without imposing additional restrictions of integration over field variables, or in the perturbation theory. The functional $\beta S_{\text{eff}}(\varphi, \lambda)$ depends on the distribution of the field variables φ and the absolute chemical activity ξ. The field variable φ contains information similar to the original partition function with summation over occupation numbers, i.e., complete information on probable states of the system. Now, the saddle-point method may be employed to find the asymptotic value of the partition function $\Omega_{E,N}$ for $N \to \infty$. The dominant contribution is by the states that satisfy the extreme condition for the functional. The particle distribution is determined by the saddle-point solutions of the equation

$$\frac{\delta S_{\text{eff}}}{\delta \xi} = \frac{\delta S_{\text{eff}}}{\delta \varphi} = 0, \tag{6.10}$$

where the distribution of particles may be either spatially inhomogeneous or not. The existence of solutions that correspond to the finite entropy $S_{\text{eff}}(\varphi, \lambda)$ while the volume of the system tends to infinity implies that such solutions could be thermodynamically stable. The above set of equations, in principle, solves the many-particle problem in the thermodynamic limiting case. The spatially inhomogeneous solution of this equation corresponds to the distribution of interacting particles. Such inhomogeneous behavior is associated with the nature and intensity of interaction. In other words, the accumulation of particles within a finite spatial region (formation of a cluster) influences the spatial distribution of the fields and the activity. The inverse matrix $\omega_{ss'}^{-1}$ of the interaction $\omega_{ss'} = \omega(|r_s - r_{s'}|)$ in the continuum limit should be considered with the operator [14], i.e.,

$$\omega_{rr'}^{-1} = \delta_{rr'}\widehat{L_{r'}} = -\frac{1}{4\pi G m^2}\Delta, \tag{6.11}$$

where m is the particle mass and Δ is the Laplace operator. With the accuracy up to the surface term, in the continuum case, the effective entropy takes the form

$$S_{\text{eff}}(\varphi, \xi) = \int dV \left\{ \frac{1}{8\pi Gm^2\beta}(\nabla\varphi)^2 + \delta \sum_p \ln\left(1 - \delta\xi e^{\varphi - \beta\varepsilon_p}\right) \right\}$$

$$+ N\ln\xi - \beta E. \tag{6.12}$$

As was shown before [1], in all the cases for high temperatures $\xi \leq 1$, we may make use of the classical Boltzmann statistics and expansion

$$\sum_p \ln\left(1 + \xi e^{\varphi - \beta\sum_p \varepsilon_p}\right) \approx \xi e^{\varphi - \beta\varepsilon_p} + \cdots.$$

The integration over the momentum and coordinates should be performed with regard for the cell volume $(2\pi\hbar)^3$ in the phase space of individual states [3]. After integration over the momentum, the effective entropy may be presented in the general form given by

$$S_{\text{eff}}(\varphi, \xi) = \int dV \left\{ \frac{1}{8\pi Gm^2\beta}(\nabla\varphi)^2 - \xi A e^\varphi \right\} + N\ln\xi - \beta E, \tag{6.13}$$

where

$$A \equiv \left(\frac{2\pi m}{\beta\hbar^2}\right)^{3/2}.$$

The microcanonical partition function thus obtained may be applied in the calculation of any thermodynamic relation for the self-gravitating system. In the most general case, the problem cannot be solved completely, but for individual cases, it is possible to obtain all thermodynamic characteristics of the system. In what follows we consider some special cases.

First of all, we consider a system of noninteracting particles, when $\varphi = 0$. The effective entropy for this case may be written in a simple form

$$S_{\text{eff}}(\varphi, \xi) = - \int dV \xi \left(\frac{2\pi m}{\beta \hbar^2} \right)^{3/2} + N \ln \xi - \beta E. \qquad (6.14)$$

The extreme condition

$$\frac{\delta S_{\text{eff}}}{\delta \xi} = \frac{\delta S_{\text{eff}}}{\delta \beta} = 0$$

reduces to two equations

$$V \xi \left(\frac{2\pi m}{\beta \hbar^2} \right)^{3/2} = N \qquad (6.15)$$

and

$$V \xi \left(\frac{2\pi m}{\beta \hbar^2} \right)^{3/2} = \frac{2}{3} \beta E. \qquad (6.16)$$

These equations directly determine the chemical activity

$$\xi = \frac{N}{V} \left(\frac{2\pi m}{\beta \hbar^2} \right)^{-3/2} \qquad (6.17)$$

and yield the well-known relation between the fixed energy and number of particles in the system and the reciprocal temperature, i.e.,

$$\beta = \frac{3}{2} \frac{N}{E} \qquad (6.18)$$

or

$$\frac{3}{2} N k T = E.$$

Substituting the coefficient obtained previously into the effective entropy, we obtain the ordinary expression for the entropy, i.e.,

$$S_{E,N} = \ln \frac{N!}{V} \left(\frac{4\pi m E}{3 N \hbar^2} \right)^{3/2} + \frac{3N}{2}. \qquad (6.19)$$

The microcanonical partition function for fixed total energy and number of particles may be written in the form

$$\Omega_{E,N} = \exp\left\{-\ln\frac{N!}{V}\left(\frac{3N\hbar^2}{4\pi mE}\right)^{3/2} + \frac{3N}{2}\right\} \tag{6.20}$$

that reproduces the well-known presentation of the microcanonical partition function for the ideal Boltzmann gas, i.e.,

$$\Omega_{E,N} = \frac{V^N}{N!}\left\{\frac{4\pi mEe^N}{3N\hbar^2}\right\}^{3N/2}. \tag{6.21}$$

In the case of a homogeneous particle distribution $\nabla\varphi = 0$ with the average interparticle distance $l = (V/N)^{1/3}$, we may introduce the average value of the potential

$$\varphi_h = \frac{3Gm^2N}{2E}\left(\frac{N}{V}\right)^{1/3} \tag{6.22}$$

and thus obtain the ordinary entropy in the form

$$S_{E,N} = \ln\frac{N!}{Ve^{\varphi_h}}\left(\frac{4\pi mE}{3N\hbar^2}\right)^{3/2} + \frac{3N}{2}. \tag{6.23}$$

The microcanonical partition function then reduces to

$$\Omega_{E,N} = \frac{V^N}{N!}e^{-N\tilde{\varphi}}\left\{\frac{4\pi mEe^N}{3N\hbar^2}\right\}^{3N/2}. \tag{6.24}$$

The microcanonical partition function solves the problem of determining of the thermodynamic properties of a self-gravitating system with homogeneous distribution of particles. From the expression for the entropy it is possible to find the pressure in a self-gravitating system using the homogeneous distribution of particles as

$$P_h = \frac{N}{\beta Ve^{\varphi_h}},$$

or the equation of state in the form

$$P_h V_h = \frac{2E}{3e^{\varphi_h}}. \tag{6.25}$$

In the general case, however, distributions of particles in self-gravitating systems are inhomogeneous. Inhomogeneous distribution of particles motivates the long-range gravitation interaction. Such systems are unstable, and in the final state, are divided into several finite-size clusters. The next task is to develop an acceptable method for determining the partition function and allowing the inhomogeneous distribution of finite-size clusters. We start from the case when the homogeneous distribution of particles decomposes into n equal clusters with the average volume V_c. Within this volume, there exists an inhomogeneous distribution of particles with the nonzero field variable, and outside of this volume, the field variable is equal to zero because there are no particles.

In this case the effective entropy may be rewritten in the form

$$S_{\text{eff}}(\varphi, \xi) = n \int_0^{V_c} dV \left\{ \frac{1}{4r_m} (\nabla \varphi)^2 - \xi A e^\varphi \right\}$$

$$- \xi A (V - nV_c) + N \ln \xi - \beta E, \tag{6.26}$$

where, as before,

$$r_m = 2\pi G m^2 \beta \quad \text{and} \quad A = \left(\frac{2\pi m}{\beta h^2} \right)^{3/2}.$$

The minimization of the effective entropy with respect to the field variables yields the saddle-point solutions of the equation inside the cluster given by

$$\frac{1}{2r_m} \Delta \varphi + \xi A e^\varphi = 0 \tag{6.27}$$

and the normalization condition yields

$$n \int_0^{V_c} dV \xi A e^\varphi + \xi A (V - nV_c) = N. \tag{6.28}$$

Multiplying the first equation by $\nabla\varphi$ and making use of the relation $\Delta\varphi = \nabla(\nabla\varphi)$, we obtain the first integral of this equation, i.e.,

$$\frac{1}{4r_m}(\nabla\varphi)^2 + \xi Ae^\varphi = \Delta^2, \tag{6.29}$$

where Δ is an unknown integral of "motion" that should come from the physical conditions. In view of the relation

$$\beta\frac{dA}{d\beta} = -\frac{3}{2}A,$$

the saddle-point equation

$$\frac{\delta(S)}{\delta\beta} = 0$$

may be rewritten in the form

$$n\int_0^{V_c} dV\left\{\frac{5}{2}\xi Ae^\varphi - \Delta^2\right\} + \frac{3}{2}\xi A(V - nV_c) = \beta E. \tag{6.30}$$

If the density function is introduced in the form

$$\rho(r) \equiv \xi Ae^\varphi,$$

then the saddle-point equations may be simplified and yield the normalization condition

$$n\int dV\rho(r) + \xi A(V - nV_c) = N \tag{6.31}$$

and the equation for the energy conservation

$$n\int dV\left\{\frac{5}{2}\rho(r) - \Delta^2\right\} + \frac{3}{2}\xi A(V - nV_c) = \beta E. \tag{6.32}$$

The usual entropy may be presented in the simple form

$$S = \frac{1}{2}N - 2\beta E + N\ln\xi. \tag{6.33}$$

The entropy presentation in terms of known parameters requires determination of the unknown reciprocal temperature β and the chemical activity ξ. Solutions of the above-mentioned equation completely solve the problem of the statistical description of a self-gravitating system. In the general case, however, these solutions are unknown. An attempt to solve the problem in the general case was made in [3].

Now we consider some solutions for a self-gravitating system in the case of several equal-sized clusters forming with close spacing of particles. This condition corresponds to the final state of a self-gravitating system. The field variable inside the cluster is constant and may be presented as the potential between two close-spaced particles, i.e.,

$$\varphi = \varphi_0 = \frac{2\pi Gm^2\beta}{2R}$$

for $r - r' = 2R$, where R is the particle size. We make use of the asymptotic value of the field variable in the center of the cluster and obtain the first integral given by

$$\Delta^2 \equiv \xi A e^{\varphi_0}.$$

The initial size of the cluster may be found from simple reasoning. We may assume that in the finite case, all particles are assembled only into n clusters and regarding the finite size of a particle that occupies the volume $V_0 = (4\pi/3)R^3$, we can estimate $nV_c \simeq NV_0$. The normalization condition in this case is given by

$$n\xi A e^{\varphi_0} + \xi A(V - nV_c) = N \tag{6.34}$$

and the equation of energy conservation is given by

$$\frac{5}{2}n\xi A e^{\varphi_0} - nV_c\xi A e^{\varphi_0} + \frac{3}{2}\xi A(V - nV_c) = E. \tag{6.35}$$

From the latter relations, we obtain the chemical activity

$$\xi A = \frac{N}{V - nV_c(1 - e^{\varphi_0})} \tag{6.36}$$

and the relation

$$\beta E = \frac{3}{2}N.$$

Thus the usual entropy may be rewritten in the simple form

$$S = -N + N\ln\xi - \beta E = -N + N\ln\xi - \frac{3N}{2}. \tag{6.37}$$

Substituting the relation obtained into the effective entropy yields

$$S = -N + N \ln \frac{N}{A[V - nV_c(1 - e^{\varphi_0})]} - \frac{3N}{2}. \quad (6.38)$$

The above presentation completely solves the problem of the statistical description of a self-gravitating system with the formation of several equal-sized clusters. We make use of the relation $N - N \ln N \approx \ln N!$ and thus obtain the ordinary entropy in the form

$$S_{E,N}^{\text{inh}} = \ln \frac{V^N}{N!} \left\{ \frac{4\pi m E e^N}{3N\hbar^2[1 - (NV_0/V)(1 - e^{\varphi_0})]} \right\}^{3N/2} + \frac{3}{2}N \quad (6.39)$$

while the partition function of the microcanonical ensemble for a self-gravitating system may be presented as

$$Z_{E,N} = \frac{V^N}{N!} \left\{ \frac{4\pi m E e^N}{3N\hbar^2[1 - (NV_0/V)(1 - e^{\varphi_0})]} \right\}^{3N/2}. \quad (6.40)$$

If the gravitation interaction between particles does not occur, then $e^{\varphi_0} = 1$ and the partition function reduces to that of the ideal Boltzmann gas of hard spheres. The equation of state in the case of an inhomogeneous particle distribution (existence of several finite-sized clusters) may be given as

$$P_{\text{inh}} V_{\text{inh}} = \frac{2E}{3[1 - (NV_0/V)(1 - e^{\varphi_0})]}. \quad (6.41)$$

Within the context of the above, we may say that the entropy in the case of an inhomogeneous distribution of particles (several finite-sized clusters) is greater than the entropy for a homogeneous distribution of particles

$$S_{\text{inh}} - S_{\text{h}} = \ln \frac{e^{\varphi_h}}{[1 - (NV_0/V)(1 - e^{\varphi_0})]} \quad (6.42)$$

for

$$e^{\varphi_h} > 1 - \frac{NV_0}{V}(1 - e^{\varphi_0}).$$

This is valid for a realistic self-gravitating system, and the homogeneous distribution of particles conforms to the requirements for the equilibrium state. This is the thermodynamic reason for the formation of inhomogeneous particle distributions. The relation for the ordinary entropy of a self-gravitating system yields the equation of state, i.e.,

$$\frac{P_{\text{inh}}V_{\text{inh}}}{P_{\text{h}}V_{\text{h}}} = \frac{e^{\varphi_{\text{h}}}}{[1 - (NV_0/V)(1 - e^{\varphi_0})]} > 1. \qquad (6.43)$$

Having assumed that the pressure in both states are equal, we may conclude that the volume of the inhomogeneous distribution of particles is greater than that of the homogeneous particle distribution of the same self-gravitating system. If the domains are unlimited, then the density of states diverges as particles expand to infinity. Therefore, there is no equilibrium state in the rigorous sense. Self-gravitating systems have a tendency to disperse. This is already the case of the ordinary gas in an infinite volume. The dispersion rate is small in general and the system may be in a quasi-equilibrium state for a relatively long time.

Actually, the present equilibrium statistical description concerns only dilute structures in the self-gravitating systems though it does not describe metastable states and does not provide information on time scales and the kinetic dynamics. The partition function has no singularities for different values of the gravitation field. The problem of describing a self-gravitating system of particles could be solved by the current approach where the entropy for a finite system could be explicitly calculated. The spatial nonuniformity of particles treated as an equilibrium state might alter the activation barrier that leads to the transformation when the system tends towards a nonequilibrium state. The gravity factor could either promote or retard such transformations depending on the system and conditions concerned.

6.2   Spatial Distribution Function

Now, we consider a system of particles interacting by gravitational attraction only. For the Newtonian attraction, the inverse operator is known to be given by

$$W_{rr'}^{-1} = \frac{-1}{4\pi Gm^2}\Delta_r\delta_{rr'},$$

where G is the gravitation constant, m is the particle mass and Δ_r is the Laplace operator.

We consider the system in the usual thermodynamic limit, i.e., the number of particles $N \to \infty$ and the volume $V \to \infty$ with $\frac{N}{V}$ fixed. We take the action in the spherical coordinates and disregard the surface contribution. Then, we may write the action for the Fermi gas as

$$S_F = \frac{1}{2\beta} \int_0^V \frac{(\nabla \varphi)^2}{4\pi G m^2} dV - \frac{1}{\omega} \int_0^V dV \int d^3p$$

$$\cdot \ln \left[1 + \xi e^\varphi \exp \left(-\beta \frac{p^2}{2m} \right) \right] + (N+1) \ln \xi, \qquad (6.44)$$

or

$$S_F = \int_0^V dV \left[\frac{(\nabla \varphi)^2}{4 r_m} - \frac{1}{\lambda^3} f_{5/2} \left(\xi e^\varphi \right) \right] + (N+1) \ln \xi, \qquad (6.45)$$

where $r_m = 2\pi G m^2 \beta$, and

$$f_{5/2} \left(\xi \right) = \frac{4}{\sqrt{\pi}} \int_0^\infty dx x^2 \ln \left(1 + \xi e^{-x^2} \right)$$

or

$$f_{5/2} \left(\xi \right) = \sum_{l=1}^\infty (-1)^{l+1} \frac{\xi^l}{l^{5/2}}$$

is the special Fermi function [53, 54]. For the Bose gas, the action may be written analogously, i.e.,

$$S_B = \frac{1}{2\beta} \int_0^V \frac{(\nabla \varphi)^2}{4\pi G m^2} dV + \frac{1}{\omega} \int_0^V dV \int d^3p \ln \left[1 - \xi e^\varphi \exp \left(-\beta \frac{p^2}{2m} \right) \right]$$

$$+ \ln \left(1 - \xi \right) + (N+1) \ln \xi$$

$$= \int_0^V dV \left[\frac{(\nabla \varphi)^2}{4 r_m} - \frac{1}{\lambda^3} g_{5/2} \left(\xi e^\varphi \right) \right] + \ln \left(1 - \xi \right) + (N+1) \ln \xi, \qquad (6.46)$$

where

$$g_{5/2}(\xi) = -\frac{4}{\sqrt{\pi}} \int_0^\infty dx\, x^2 \ln\left(1 - \xi e^{-x^2}\right) = \sum_{l=1}^\infty \frac{\xi^l}{l^{5/2}}$$

is the special Bose function [53, 54]. In the case $\xi < 1$ that corresponds to the absence of the Bose condensate we have $\langle n_o \rangle / V \to 0$. Then the action of the condensate $\ln(1 - \xi)$ may be omitted.

The occurrence of the additional field φ implies that the spatial distribution function (density) may be expressed as

$$\rho = \frac{m}{(2\pi\hbar)^3} \int_0^\infty d^3 p \frac{\xi \exp\left(-\dfrac{p^2}{2mkT}\right) e^\varphi}{1 + \xi \exp\left(-\dfrac{p^2}{2mkT}\right) e^\varphi}, \tag{6.47}$$

where the expression for the integrand has been made use of. The equation for the saddle point is the Lagrange equation for functionals (6.44) and (6.46), i.e.,

$$\frac{\partial^2 \varphi}{\partial R^2} + \frac{2r_m}{\lambda^3} \frac{\partial f_{5/2}(\xi e^\varphi)}{\partial \varphi} = 0, \tag{6.48}$$

$$\frac{\partial^2 \varphi}{\partial R^2} + \frac{2r_m}{\lambda^3} \frac{\partial g_{5/2}(\xi e^\varphi)}{\partial \varphi} = 0. \tag{6.49}$$

Unfortunately, these equations have no analytical solution. But the problem is simplified in the Boltzmann limiting case.

Boltzmann limiting case

Let us consider the limiting case $\xi \to 0$ (it corresponds to high temperatures and low concentrations) using the decomposition of the special functions $f_{5/2}(\xi e^\varphi)$ and $g_{5/2}(\xi e^\varphi)$ in power series of the activity ξ. Then action (6.44) and (6.46) is reduced to one simpler action

$$S = \int_0^V \left\{ \frac{(\nabla\varphi)^2}{4r_m} - \frac{\xi}{\lambda^3} e^\varphi \right\} dV + N \ln \xi. \tag{6.50}$$

An analogous expression was obtained in [8]. However, it contains no term $N \ln \xi$ that would fix the number of particles. The equation for the saddle point of the action is given by

$$\frac{\partial^2 \varphi}{\partial R^2} + \frac{2r_m}{\lambda^3} \xi e^{\varphi} = 0. \tag{6.51}$$

In order to relate the auxiliary field φ to the density $\rho(R)$ we have to apply Eq. (6.47) and to pass to the Boltzmann limiting case $\xi \to 0$. Thus, we have

$$\rho = m \frac{\xi}{\lambda^3} e^{\varphi}. \tag{6.52}$$

Then, we may rewrite Eq. (6.51) for the field φ as an equation for the density given by

$$\frac{\partial^2 \rho}{\partial R^2} - \frac{1}{\rho}\frac{\partial \rho}{\partial R} + \frac{2r_m}{m}\rho^2 = 0. \tag{6.53}$$

This equation has a soliton solution [3]

$$\rho(R) = m \frac{\Delta^2}{r_m^3} \frac{1}{\cosh^2 \Delta(R/r_m)}, \tag{6.54}$$

where Δ is an unknown integration constant to be found somewhat later. Any soliton solution corresponds to a spatially inhomogeneous distribution of particles — a finite-size cluster. The relevant asymptotics for (6.54) is $\rho = mN/V$ for $R = D$, where D is the cluster size, and $\rho \to 0$ as $R \to \infty$. This solution describes the involvement of particles in the inhomogeneous formation of size D and the absence of particles at infinity.

In order to find the equilibrium cluster radius D, we have to write the free energy as a function of D, using the action (6.50) and to minimize it in a manner similar to [5]. However, we shall do it in a different manner.

We assume that the average energy of the gravitation interaction of two particles is lower than the average kinetic energy $\sim kT$ of a particle, and $(r_m^3 N)/V \ll 1$. This leads to the two consequences, i.e.,

(a) The activity should be given by

$$\xi = \xi_0 \left[1 + O \left(\frac{r_m^3}{V/N} \right) \right], \tag{6.55}$$

where ξ_0 is the activity for the ideal gas. Thus the activity of the system is somewhat different from the activity of the ideal gas.

(b) The density in the center is somewhat different from the average density, $\rho(0) \gtrsim mN/V$. Then, we may suppose that $\Delta^2 \gtrsim (N/V)r_m^3$.

Now we calculate the above-mentioned size D (radius of a cluster) where particles are accumulated in the above-mentioned slightly nonideal case. To do this, we calculate the potential energy of a particle $U(R)$ in the cluster using the Boltzmann distribution

$$\rho(R) = \rho(0) \exp \left(-\frac{U}{kT} \right). \tag{6.56}$$

Then from (6.54) we have

$$U(R) = -kT \ln \left[\frac{\rho(R)}{\rho(0)} \right] = -kT \ln \left[\frac{m \dfrac{\Delta^2}{r_m^3} \dfrac{1}{\cosh^2 \Delta(R)/(r_m)}}{m \dfrac{\Delta^2}{r_m^3}} \right]$$

$$= 2kT \ln \left[\cosh \left(\Delta \frac{R}{r_m} \right) \right]. \tag{6.57}$$

In view of the above assumption on gravitation and thermal energies and for small distances from the coordinate origin $R/r_m < 1$, we may write that

$$U(r) \approx 2kT\Delta^2 \frac{R^2}{r_m^2} \approx \frac{2kTr_m N}{V} R^2. \tag{6.58}$$

We see that a particle in a cluster is contained in a potential well. Then the cluster radius D may be regarded as the distance from the center of the cluster such that $U(D) = (1/2)kT$ (the factor $1/2$ corresponds to the radial degree of freedom). This means that D is the point of stop of a particle with the average kinetic energy $1/2kT$.

Solving the above equation yields the equilibrium radius of a cluster, i.e.,

$$D^2 = \frac{kTV}{4\pi Gm^2 N},$$ (6.59)

that reproduces the result of Refs. [5, 6] with the accuracy of a constant. This expression means that the equilibrium size of the cluster is determined by the balance of two forces. The first of these is gravitation that compresses the gas. It is represented by the factor Gm^2 (in the sense of the interaction constant). The decrease of the cluster size with the increase of the mean density in the system N/V is associated with closer packing of particles in the cluster caused by the increase of the gravitation energy. The second force is thermal, it produces positive pressure that is resistant to the gravitational compression. It is represented by the factor kT. During the compression of the self-gravitating system, its temperature can be constant. This balance produces equilibrium distribution in the system. That is why the dependence (6.59) is meaningful only for small changes of the temperature $\Delta T \ll T$ under the formation of the cluster, that is, in a weakly nonideal gas.

On the other hand, Eq. (6.59) may be regarded as a characteristic distance in the system, for which considerable deviation from the mean fixed density N/V is observed. Then this expression may be interpreted as Jeans length [157].

Cluster size for arbitrary temperatures

In order to calculate the radius of a cluster for any temperature we have to solve Eqs. (6.48) and (6.49) and obtain the spatial distribution function of the type (6.54). In the general case, we cannot do it via the mathematical complications mentioned above. If we suppose, however, that gas is weakly nonideal, i.e.,

$$\frac{r_m^3 N}{V} \ll 1,$$

and consider short distances from the coordinate origin, $R/r_m < 1$, where $|\varphi| \ll 1$ (see Eqs. (6.52) and (6.54)), then the problem is simplified.

For example, we consider Eq. (6.51) and expand e^φ in a series, i.e.,

$$\frac{\partial^2 \varphi}{\partial R^2} + \frac{2 r_m \xi_0}{\lambda^3} + O(\varphi) = 0, \qquad (6.60)$$

where we suppose that $\xi \approx \xi_0$, i.e., the gas is weakly nonideal. The solution and the corresponding density are given by

$$\varphi \approx -\frac{r m \xi_0}{\lambda^3} R^2 \Rightarrow \rho = m \frac{\xi}{\lambda^3} \exp\left(-\frac{r m \xi_0}{\lambda^3} R^2\right). \qquad (6.61)$$

Then, using distribution (6.56), we obtain the potential energy of a particle in a cluster as given by

$$U(R) = -kT \ln\left[\exp\left(-\frac{r m \xi_0}{\lambda^3} R^2\right)\right] = kT \frac{r m \xi_0}{\lambda^3} R^2, \qquad (6.62)$$

that could be obtained before (6.58). Then, using the condition $U(D) = kT/2$ and the expression for the Boltzmann activity $\xi_0 = \lambda^3 N/V$, we obtain

$$kT \frac{r m \xi_0}{\lambda^3} D^2 = \frac{1}{2} kT \Longrightarrow D^2 = \frac{V kT}{4\pi G m^2 N}. \qquad (6.63)$$

This result reproduces Eq. (6.59).

Now let us calculate the cluster radius for any temperature using Eq. (6.48). For the Bose system, we are interested in the first quantum correction only because function $g_{5/2}$ has a singularity for low temperatures (Bose–Einstein condensation). This case was considered in [5]. We rewrite Eq. (6.48) as

$$\frac{\partial^2 \varphi}{\partial R^2} + \frac{2 r m}{\lambda^3} \left(\frac{\partial f_{5/2}\left(\xi_0 e^\varphi\right)}{\partial \varphi}\right)_{\varphi=0} + O(\varphi). \qquad (6.64)$$

Analogously, the scale of the spatial inhomogeneity is determined by the term before the zero degree of the field φ. However, the above calculation, Eqs. (6.61)–(6.63), is disputable in the general case. For example, suppose we wish to calculate the cluster radius for the limiting case $T \to 0$. Then, we have to use the spatial distribution function given by

$$\rho \approx m \int_0^{p_F} 4\pi p^2 dp \left[1 - \frac{1}{\xi_0} \exp\left(\frac{p^2}{2mkT}\right) e^{-\varphi} \right], \qquad (6.65)$$

because

$$\exp\left(\frac{\varepsilon_F - \varepsilon}{kT}\right) \gg 1, \quad \text{and} \quad \xi_0 = \exp\left(\frac{\varepsilon_F}{kT}\right).$$

However, the integrand cannot be integrated in elementary functions. Hence, we have to find some other approach.

For an approximate solution of this problem we propose to use the particle energy of the form

$$U(R) = -kT \ln \left[\frac{\rho(R)}{\rho(0)}\right] = -kT \ln e^\varphi = -kT\varphi, \qquad (6.66)$$

where the distribution (6.52) has been used, and the cluster radius D is given by

$$U(D) = \frac{1}{2}kT + \frac{3}{5}\varepsilon_F, \qquad (6.67)$$

with ε_F being the Fermi energy. The term $kT/2$ is the particle kinetic energy associated with the thermal motion and $3\varepsilon_F/5$ is the average particle kinetic energy in the degenerated Fermi gas (because all particles cannot be at the level $\varepsilon = 0$ due to Pauli principle). The term $3\varepsilon_F/5$ should be treated as the repulsive "statistical potential" [53].

The scale of the spatial inhomogeneity is determined by the term before the zero degree of the field φ. Then the cluster radius is determined by a set of two equations (the second one is Eq. (5.19) for the activity ξ_0) and

$$kT\frac{2r_m}{\lambda^3}\left(\frac{\partial f_{5/2}(\xi_0 e^\varphi)}{\partial \varphi}\right)_{\varphi=0} D^2 = \frac{1}{2}kT + \frac{3}{5}\varepsilon_F,$$

$$\frac{\lambda^3 N}{V} = f_{3/2}(\xi_0). \qquad (6.68)$$

Now, we consider some limiting cases. In the limiting case of high temperatures (for Bose statistics, we should employ the expansion of the function $g_{5/2}$ rather than $f_{5/2}$) we suppose that $kT \gg \varepsilon_F$. Then we calculate the first quantum correction in the expansion (5.19) and thus obtain

$$D^2 = \frac{kTV}{4\pi Gm^2 N} \pm \frac{1}{2^{5/2}} \frac{\lambda^3}{r_m}, \tag{6.69}$$

where "+" corresponds to the Fermi gas and "−" corresponds to the Bose gas, and $\lambda^3/r_m^3 \ll 1$. The meaning of the zero approach was explained in Sec. 6.2. The meaning of the first quantum correction may be explained as follows.

The so-called "statistical potential" of particle interaction is known to occur in quantum gases [53] — it is repulsion for fermions and attraction for bosons. The phenomenon is associated with the symmetry of particle wave functions. Then the equilibrium size is determined by three energies — gravitation (the "constant of interaction" is r_m), thermal kT, and the above-mentioned "statistical potential" (the "constant of interaction" is λ^3). That is, in the Fermi case, the quantum energy is resistant to the compression caused by the gravitation force (as a small addend to the thermal energy), and in the Bose case, the quantum energy compresses the gas along with the gravitation energy (as a small addend). This explains the signs ("+" or "−") of the small corrections to the cluster radius.

In the limiting case of low temperatures, we suppose that $kT \ll \varepsilon_F$, calculate the first temperature correction using kT as a small addend, and thus obtain

$$D^2 \approx 10^{-3} \frac{\hbar^2}{Gm^3} \left(\frac{V}{N}\right)^{1/3} \left[1 + \frac{5}{6}\frac{kT}{\varepsilon_F}\right]. \tag{6.70}$$

We see that the equilibrium size is determined by two energies in the zero approach — the gravitation energy and the above-mentioned repulsion statistical potential (the "constant of interaction" is $\hbar$). That is, in the Fermi case, the quantum energy that is proportional to the concentration that decreases from the cluster center to infinity, is resistant to the gravitation compression under the action of the gravitation force. The small correction kT/ε_F means that the above-mentioned thermal mechanism of cluster formation (as a small addend to the Fermi statistical repulsion) occurs for $T \neq 0$.

Finite self-gravitating system

So far, we have considered the self-gravitating system in the limiting case $N \to \infty$, $V \to \infty$, for fixed N/V. Realistic systems, however, are known to have finite volumes and numbers of particles. In the case of a system with gravitation interaction between the particles, the situation should be changed cardinally because the range of such interaction is longer than the size of the system.

So Eq. (6.48) has a soliton solution for any thermodynamic conditions, that is, the cluster thus formed exists permanently. The situation is changed in a finite system where N and V are fixed. Let us consider this case in more detail.

Suppose the system is a gravitating gas of mass mN that occupies the volume V. This gas produces the gravitation field $\Delta\phi = 4\pi G\rho$, where $\rho \simeq mN/V$ and $V \sim R_{max}^3$. For a calibration such that the field $\phi = 0$ on the boundary of the system, the energy of a particle in the center of the system (the depth of the potential well) is given by

$$U_0 = m\phi \sim -mG\rho R_{max}^2. \tag{6.71}$$

On the other hand, the thermal energy of a particle is $\sim kT$. If the thermal energy of a particle is lower than its maximum gravitation energy (in the center of the system), then the gas has negative pressure and tends to compress to a point. Contrarily, if the thermal energy of a particle is higher than its maximum gravitation energy (by the module), then the gas has positive pressure and tends to spread to infinity. During compression of the gas, its thermal energy increases because the total energy of the system is constant. Thus, the equilibrium function of spatial distribution (6.54) is established.

Then, the condition of cluster formation is determined by the correlation between the above-mentioned energies as given by

$$mG\rho R_{max}^2 \gtrsim kT \Rightarrow \frac{Gm^2 N}{kT R_{max}}. \tag{6.72}$$

If we rise this equation to the third power, then it reproduces Eq. (6.59) with an accuracy of a constant.

If $N \to \infty$ and $V \to \infty$, then $U_0 \to \infty$. This means that the thermal energy is lower than the gravitation energy in the center of the system always (in this case, any point may be chosen for the center of the system), hence such self-gravitating gas collapses for any temperature.

Similarly, in the degenerated case we have

$$mG\rho R_{\max}^2 \gtrsim \varepsilon_F \Rightarrow \frac{Gm^3 N^{1/3} R_{\max}}{\hbar^2} \gtrsim 1. \tag{6.73}$$

The same equation may be obtained from the criterion $D \gtrsim R_{\max}$ within the context of Eqs. (6.69) and (6.70).

6.3   Inhomogeneity of Self-Gravitating Systems

Statistical approach

It is not difficult to notice that replacing the interaction constant Gm^2 in formulas (6.59) and (6.70) for cluster radii by the electric interaction constant q^2 yields the Debye radius and the Thomas-Fermi radius in a plasma, respectively. Thus an idea arises that some analogy exists of processes and formations in the two above-mentioned cases. Let us explain this coincidence.

Suppose N_+ particles repulse by the Coulomb law with the interaction constant Gm^2, i.e.,

$$U_{ij} = \frac{Gm^2}{|\mathbf{R}_i - \mathbf{R}_j|},$$

(unlike the Newtonian attraction considered above). In order to consider the stability of such a system we assume that these particles are immersed in the environment of the opposite sign (with charges $-m$) so that the neutrality condition is satisfied, $N_+ = N_- = N$, but the current is produced by N_+ particles only. In other words, we consider a plasma with the interaction constant Gm^2 and charge carriers of one of the signs being immovable. Let us find the spatial distribution function for a system where $N \to \infty, V \to \infty$, but N/V is fixed.

Let a particle with charge m be in the coordinate origin. It polarizes the gas of particles with like charges (the environment with the charges of the opposite sign is immovable). For particle repulsion, the action should be written as

$$S = \int_0^V \left[\frac{(\nabla \psi)^2}{4 r_m} - \frac{1}{\lambda^3} f_{5/2}(\xi \cos \psi) \right] dV + (N+1) \ln \xi, \qquad (6.74)$$

and the relevant Lagrange equation is given by

$$\frac{d^2 \psi}{dR^2} + \frac{2 r_m^3}{\lambda^3} \frac{\partial f_{5/2}(\xi \cos \psi)}{\partial \psi} = 0. \qquad (6.75)$$

In the Boltzmann limiting case, $\xi \to 0$, we may write this expression and the spatial distribution function as

$$S = \int_0^V \left[\frac{(\nabla \psi)^2}{4 r_m} - \frac{\xi}{\lambda^3} \cos \psi \right] dV + (N+1) \ln \xi \qquad (6.76)$$

$$\rho(R) = m \frac{\xi}{\lambda^3} \cos \psi \qquad (6.77)$$

with the relevant Lagrange equation

$$\frac{d^2 \psi}{dR^2} - 2 \frac{r_m^3}{\lambda^3} \xi \sin \psi = 0. \qquad (6.78)$$

We assume that $\xi = \xi_0 = (\lambda^3 N)/V$ and seek for the solution in the domain $\psi(R) \in [0, \pi/2]$ where $R \in [0, +\infty]$, that is, with asymptotics $\psi(0) = \pi/2$ and $\psi(+\infty) = 0$. The first asymptotics corresponds to the absence of particles in the coordinate origin, because in order to penetrate the coordinate origin, a particle should have infinitely high kinetic energy. The second asymptotics means that for the concentration equal to the average concentration at large distances $\rho(R \to \infty) = mN/V$. In other words, the spatial distribution function should have a fall. Such formation is usually called a polaron, unlike a cluster — a system with attraction where the distribution has a hump. Then the solution of Eq. (6.77) is given by

$$\tan \frac{\psi}{4} = \tan \frac{\pi}{8} \exp\left(-\sqrt{\frac{2r_m\xi_0}{\lambda^3}}R\right), \tag{6.79}$$

and the spatial distribution function is given by

$$\rho(R) = \frac{mN}{V} \cos\left\{4 \arctan\left[\exp\left(-\sqrt{\frac{2r_m\xi_0}{\lambda^3}}R\right)\tan\frac{\pi}{8}\right]\right\}. \tag{6.80}$$

This distribution is determined by the balance of two energies, i.e., the Coulomb energy preventing particle penetration into the coordinate origin and the thermal (average particle kinetic) energy. Then the polaron radius may be found as the point of stop, $U(D_p) \sim kT$. The function $U(R)$ may be found from the Boltzmann distribution

$$\rho(R) = \rho(R \to \infty) \exp\left(-\frac{U}{kT}\right). \tag{6.81}$$

Then using (6.79) we have

$$U = -\ln\left\{\cos\left[4 \arctan\left(\exp\left(\sqrt{-2\frac{r_m\xi_0}{\lambda^3}}R\right)\tan\frac{\pi}{8}\right)\right]\right\}. \tag{6.82}$$

In order to satisfy the condition $U(D_p) \sim kT$ we have to assume that

$$D_p^2 \sim \frac{\lambda^3}{r_m\xi_0} = \frac{kTV}{Gm^2N}. \tag{6.83}$$

This expression for the polaron radius is similar to the cluster radius (6.59) with the accuracy of a constant factor. This value may be interpreted as the scale on which the neutrality of the gas is violated.

The coincidence of the sizes D and D_p may be explained as follows. We have seen that spatial distributions in both cases are determined by the two energies only — the Coulomb-type energy and the thermal energy. Then the formation mechanisms of a cluster and a polaron are similar — the balance of the two energies. These differ by the effects only: in the gravitating gas they produce the spatial distribution of the hump type and in the case of repulsing particles we have a fall. In the degenerated case, the situation is analogous, just with the Fermi energy occurring against the thermal energy.

Hydrodynamical approach

In order to explain the coincidence of the radii of a cluster D and a polaron D_p we can employ the hydrodynamical approach that makes it possible to consider some general properties of a system with Coulomb-type interaction from somewhat different point of view. Suppose a volume "charge" ρ appears in the media under consideration. Any change is described by the continuity equation, i.e.,

$$\frac{\partial \rho}{\partial t} = -\operatorname{div} \mathbf{j}, \tag{6.84}$$

where $\mathbf{j}$ is the flow of mass or charge that is given by

$$\mathbf{j} = \langle n \rangle m \mathbf{v} = m \frac{N}{V} \mathbf{v}, \tag{6.85}$$

where $\mathbf{v}$ is the particle velocity. The motion of particles is described by the equation:

$$m \frac{d\mathbf{v}}{dt} = m\mathbf{E}, \tag{6.86}$$

where $\mathbf{E}$ is the field produced by the remaining particles at the point where the particle of mass m is located. Then the *linearized* equation of motion for the fluctuations of ρ is given by

$$\frac{\partial^2 \rho}{\partial t^2} = -\frac{Nm}{V} \operatorname{div} \mathbf{E}. \tag{6.87}$$

The equations for the field $\mathbf{E}$ for gravitating particles and for repulsing particles differ only by the sign of the right-hand-side part, i.e.,

$$\operatorname{div} \mathbf{E} = -4\pi G \rho \tag{6.88}$$

$$\operatorname{div} \mathbf{E} = 4\pi G \rho \tag{6.89}$$

for gravitating and repulsive particles, respectively. Then, we may write the final equations of motion for gravitating and repulsing particles, respectively, as

$$\frac{\partial^2 \rho}{\partial t^2} = \frac{4\pi N m}{V} \tag{6.90}$$

$$\frac{\partial^2 \rho}{\partial t^2} = -\frac{4\pi N m}{V}. \tag{6.91}$$

We may separate the solutions of these equations in the form

$$\rho \sim \exp(-\omega t) \tag{6.92}$$

$$\rho \sim \exp(i\omega t), \tag{6.93}$$

where

$$\omega^2 = \frac{4\pi N m}{V}.$$

That is, in the case of gravitating particles, we see that the system is relaxing towards some homogeneous distribution. In the case of repulsing particles, we observe oscillations in the system (plasma oscillations). We see, however, that both these processes are characterized by the same time scale $\tau \sim 1/\omega$. Hence, we may suppose that the relevant spatial scale is the distance passed by a particle during its thermal motion for time $1/\omega$, i.e.,

$$d \sim \frac{\langle v \rangle}{\omega} = \sqrt{\frac{kT}{m}} / \omega = \sqrt{\frac{kTV}{4\pi G m^2 N}}. \tag{6.94}$$

This expression reproduces the radii D and D_p of gravitating and repulsing particles, respectively. Thus, both systems are characterized by similar master equations. In the more general quantum case the reasoning is analogous.

We have already mentioned that the partition function of a self-gravitating system diverges. Nevertheless, there occur states for which the equilibrium spatial distribution function is the solution of Eq. (6.48). This means that our system disintegrates (collapses) to infinite number of clusters of size D_0 each because the number of particles in the system tends to infinity, $N \to \infty$, while the number of particles in the cluster is finite. This multitude is a self-gravitating system too and may collapse. Then the process is repeated again. In other words, the free energy of the system has no absolute minimum and each state of the self-gravitating system is analogous to the false vacuum in the field theory [24]. To estimate the relaxation times, we may apply Eq. (6.94) for the time scale. So, for each act of collapse, we have to substitute $m \to nm$ and $N \to N/m$, where n is the average number of particles in a cluster. Then the time scale $1/\omega$ is constant for each cluster formation (collapse). In other words, the state of the system with the spatial distribution function (6.54) may be referred to as metastable.

6.4 Conditions for the Gravothermal Catastrophe

Infinite system

Now we consider a system of particles interacting by gravitational attraction only. For the Newtonian attraction, the inverse operator is known to be

$$W_{rr'}^{-1} = \frac{-1}{4\pi Gm^2} \Delta_r \delta_{rr'},$$

where G is the gravitation constant, m is the particle mass, and Δ_r is the Laplace operator.

We consider the system in the usual thermodynamical limiting case, i.e., the number of particles $N \to \infty$ and the volume $V \to \infty$ with N/V fixed. Then, going to the Boltzmann limit, we may write the action in the form

$$S = \int_0^V \left\{ \frac{1}{2\beta} \frac{(\nabla \varphi)^2}{4\pi g^2} - \frac{\xi}{\lambda^3} e^\varphi \right\} dV + N \ln \xi. \qquad (6.95)$$

It is not difficult to notice that the description of the self-interacting gas in thermal equilibrium is similar to the field theory of a single scalar field $\varphi(r)$ with exponential self-action. A relevant expression was obtained in [8], however, it does not contain the entropy term.

We introduce a dimensionless quantity $r = R/r_m$ instead of R, a new variable $\sigma = \exp(\varphi/2)$, and denote $\alpha^2 \equiv r_m^3/\lambda^3$, with $r_m = 2\pi Gm^2\beta$. Then the action (in spherical coordinates) may be written in terms of the new variable, we have

$$S = 4\pi \int_0^\infty \left[\left(\frac{1}{\sigma} \frac{\partial \sigma}{\partial r} \right)^2 - \alpha^2 \xi \sigma^2 \right] r^2 dr + N \ln \xi. \tag{6.96}$$

The meaning of the additional field φ or its equivalent σ is that the spatial distribution function (density) may be expressed as, i.e.,

$$\rho(r) = m \frac{\xi}{\lambda^3} e^\varphi \equiv m \frac{\xi}{\lambda^3} \sigma^2. \tag{6.97}$$

The equation for the saddle point is the Lagrange equation for the function. Here, it is given by

$$\frac{\partial^2 \sigma}{\partial r^2} - \frac{1}{\sigma} \left(\frac{\partial \sigma}{\partial r} \right)^2 + \xi \alpha^2 \sigma^3 = 0. \tag{6.98}$$

This equation has a soliton solution [3, 5, 6].

$$\sigma = \frac{\Delta}{\sqrt{\xi} \alpha} \frac{1}{\cosh \Delta r}, \tag{6.99}$$

where Δ is the integration constant. Any soliton solution corresponds to a spatially inhomogeneous distribution of particles — a finite-size cluster. The relevant asymptotics is $\sigma^2 = 1$ for $r = d$, where d is the cluster size, and $\sigma \to 0$ as $r \to \infty$. This solution describes the involvement of particles in the inhomogeneous formation of size d and nonoccurrence of particles at infinity. This spatial distribution is compressed to the line $\sigma = 1 \Rightarrow \varphi = 0$ as far as $T \to \infty$, $N/V \to 0$. Hence, the field $\varphi = 0$ corresponds to the spatially homogeneous distribution in a self-gravitating gas in the above-mentioned limiting case.

We substitute the solution in the action

$$S = 4\pi \int_0^d \left(\Delta^2 - 2\xi \alpha^2 \sigma^2 \right) r^2 dr + N \ln \xi. \tag{6.100}$$

To perform integration, we make use of the expansion of $1/\cosh x \approx 1 - x^2/2$ in power series of $x \equiv \Delta d \ll 1$. Thus, we have

$$S = -V \frac{\Delta^2}{\alpha^2 \lambda^3} + N \ln \xi, \tag{6.101}$$

where Δ^2 is found from the asymptotics

$$1 = \frac{\Delta^2}{\xi \alpha^2} \left[1 - \Delta^2 d^2 \right] \implies \Delta^2 \approx \xi \alpha^2 + \xi^2 d^2 \alpha^4.$$

As a result, we have

$$S = -\frac{V}{\lambda^3} \xi + N \ln \xi - \frac{V}{\lambda^3} \xi^2 d^2 \alpha^2. \tag{6.102}$$

Assuming that $\lambda^3/V \gg \lambda^6/V^2$, $\xi_0 \gg \xi_G$, where ξ_0 and ξ_G are the activities of the ideal and gravitating gases, respectively, we find the activity ξ from the saddle point equation $\partial S/\partial \xi = 0$ to be given by

$$\xi \approx \frac{\lambda^3 N}{V} - \frac{2 d^2 \alpha^2 \lambda^6 N^2}{V^2} \equiv \xi_0 + \xi_G. \tag{6.103}$$

Then integrating in the saddle point, we obtain the partition function as

$$Z_N = Z_N^0 \exp \left[\frac{V}{\lambda^3} \xi_G - N \ln \left(1 + \frac{\xi_G}{\xi_0} \right) + \frac{V}{\lambda^3} \xi^2 d^2 \alpha^2 \right], \tag{6.104}$$

where Z_N^0 is the partition function of the ideal gas. The knowledge of the latter makes it possible to find the free energy of the system. Thus, we have

$$F = F_0 - kT \left[\frac{V}{\lambda^3} \xi_G - N \ln \left(1 + \frac{\xi_G}{\xi_0} \right) + \frac{V}{\lambda^3} \xi^2 d^2 \alpha^2 \right], \tag{6.105}$$

where F_0 is the free energy of the ideal gas. Minimizing with respect to the cluster size $d = D/r_m$ yields

$$\frac{\partial F}{\partial d} = -kT \frac{d \alpha^2 \lambda^3 2 N^2}{V} \times \left[1 - 4 \frac{d^2 \alpha^2 \lambda^3 N}{V} \right] = 0, \tag{6.106}$$

and thus we obtain the optimum radius of the cluster to be given by

$$d_0^2 = \frac{V}{4 N r_m^3} \tag{6.107}$$

or in the dimensionless variables

$$D_0^2 = \frac{1}{4} \frac{V kT}{2 \pi G m^2 N}. \tag{6.108}$$

The decrease of the cluster size with the increase of the average density N/V is associated with closer packing of particles in the cluster due to the increase of the gravitation energy. The increase of the cluster size with temperature is associated with less close packing of particles due to the resistance of the thermal-motion energy $(\sim kT)$ to the gravitation energy. Such a situation is realized due to the long-range attraction $(\sim 1/R)$ of the gravitation interaction.

Given that the number of particles in the system is infinite while the number of particles in the cluster is finite, this means that our system disintegrates into infinite multitude of clusters of size D_0 each. This multitude makes a self-gravitating system too and then collapses. The process is then repeated again. In other words, the free energy of such a system has no absolute minimum and each state of the self-gravitating system is analogous to the false vacuum in the field theory [24].

Finite system

Eq. (6.98) has a soliton solution under any thermodynamic conditions,

$$\alpha^2 \xi \equiv \frac{r_m^3 N}{V},$$

when $N \to \infty$, $V \to \infty$, but N/V is fixed. This means that the gas collapses for any initial condition, and the collapse does not occur only for infinitely high temperatures. The dependence of the cluster size on the thermodynamic conditions confirms to the properties of the gravitation interaction. However, in the papers [8] we see that there may exist such finite temperature that the collapse cannot occur if the temperature is higher than this value. Now we consider the reasons for these contradictions and the nature of the gravothermal catastrophe.

Let the number of particles N be very large though finite, the average density N/V be fixed. Then, we may write the normalization condition for the spatial distribution function (6.97) as

$$\int_0^V \rho(r)d^3r = \frac{mN}{r_m^3} \quad \text{or} \quad \int_0^V \sigma^2 d^3r = \frac{V}{r_m^3}. \tag{6.109}$$

This equation may be regarded as the condition imposed on the solution of Eq. (6.98). We have to find the field σ that minimizes the action under the condition (6.44). Then the integrand of (6.44) multiplied by the undetermined Lagrange multiplier should be added to the functional in

$$\left(\frac{1}{\sigma}\frac{d\sigma}{dr}\right)^2 - (\xi\alpha^2 - \chi)\sigma^2, \tag{6.110}$$

where χ is the undetermined Lagrange multiplier. The corresponding action has a minimum for the solution of the equation

$$\frac{d^2\sigma}{dr^2} - \frac{1}{\sigma}\left(\frac{d\sigma}{dr}\right)^2 + (\xi\alpha^2 - \chi)\sigma^3 = 0. \tag{6.111}$$

The multiplier χ is a function of N such that $\lim_{N\to\infty}\chi = 0$ because Eq. (6.48) should be reduced to Eq. (6.98) in the limiting case $N \to \infty$. The soliton solution is realized provided the inequality

$$\xi\alpha^2 - \chi \geq 0 \Rightarrow \frac{r_m^3 N}{V\chi} \geq 1, \tag{6.112}$$

is satisfied. The relevant equality determines the point (temperature and concentration) of the transition between, as it will be shown below, spatially homogeneous and inhomogeneous distributions, i.e., collapse.

The multiplier χ may be found by solving the system of three equations. Two of these are determined by the asymptotics of the solution of Eq. (6.48) and the third one is the normalization condition (6.44). However, we may take a simpler way. We compare two actions *at the saddle point* $\tilde{\xi}$. One of these is the action of the collapsed gas. It may be obtained from the normalization condition (6.44), we have

$$S_d = 4\pi \int_0^d \left(\Delta^2 - 2\xi\alpha^2\sigma^2\right)r^2 dr + N\ln\xi$$

$$= \Delta^2\frac{V}{r_m^3} - 2\xi\alpha^2\frac{V}{r_m^3} + N\ln\xi. \tag{6.113}$$

The second action corresponds to the spatially homogeneous distribution with $\varphi \equiv 0$, i.e.,

$$S_\infty = -\frac{V}{\lambda^3}\xi + N\ln\xi. \qquad (6.114)$$

The saddle points for both actions are determined by the equations

$$\frac{\partial S_d}{\partial \xi} = 0 \Rightarrow \tilde{\xi}_d = \frac{1}{2}\frac{\lambda^3 N}{V} \qquad (6.115)$$

$$\frac{\partial S_\infty}{\partial \xi} = 0 \Rightarrow \tilde{\xi}_\infty = \frac{\lambda^3 N}{V}. \qquad (6.116)$$

It is not difficult to notice that thermodynamic conditions determine the saddle point for both actions and in the case of collapse, the action for the collapsed gas should be less than that for the spatially homogeneous distribution on the saddle point, i.e.,

$$S_d(\tilde{\xi}_d) \le S_\infty(\tilde{\xi}_\infty)$$

$$\Downarrow$$

$$\Delta^2\frac{V}{r_m^3} - 2\tilde{\xi}_d\alpha^2\frac{V}{r_m^3} + N\ln\tilde{\xi}_d \le -\frac{V}{\lambda^3}\tilde{\xi}_\infty + N\ln\tilde{\xi}_\infty, \qquad (6.117)$$

where the equality is attained at the point of collapse. The inequality in (6.50) may be reduced to the form

$$\frac{r_m^3 N^3}{V\Delta^2(2+\ln 2)} \ge 1. \qquad (6.118)$$

Having compared this expression to Eq. (6.49) we find that

$$\chi \equiv \Delta^2(2+\ln 2).$$

We integrate over the whole space with the use of the normalization (6.44) and thus obtain

$$\Delta = \frac{\pi^3}{3N}.$$

Then for temperatures or concentrations for which the inequality

$$\frac{r_m^3 N^3}{V} \ge \left(\frac{\pi^3}{3}\right)^2(2+\ln 2) \qquad (6.119)$$

is satisfied, the gravitating gas is in the collapsed state. The same parameter is mentioned in articles [8] and reproduces the Jeans criterion of the gravitation instability. The process is associated with the gravitation energy increase that accompanies the increase of the average concentration N/V and the decrease of the thermal energy that accompanies the decrease of temperature.

For $r_m^3 N^3 / V\chi < 1$ the spatial distribution is equal to

$$\sigma = \frac{\Delta}{\sqrt{\xi}\alpha} / \sinh(\Delta r).$$

In this case, the corresponding action is less than the action for the homogeneous spatial distribution

$$\Delta^2 \frac{V}{r_m^3} + 2\xi\alpha^2 \frac{V}{r_m^3} + N \ln \xi > -\frac{V}{\lambda^3} + N \ln \xi \tag{6.120}$$

so far as

$$\Delta^2 \frac{V}{r_m^3} + 2N + N \ln \xi \geq -N + N \ln \xi. \tag{6.121}$$

Therefore, the homogeneous spatial distribution is realized under this condition.

It is not difficult to notice that we can dispense with the Lagrange multiplier for the computation of condition (6.51). This approach, however, reveals the origin of the two types of solution of the present equation – the one associated with the collapsed gas and the other one associated with the spatially homogeneous distribution.

The equality (6.51) determines the temperature after which further cooling is accompanied by the collapse. This temperature is given by

$$kT_m = \left(\frac{36}{\pi^3(2 + \ln 2)} \frac{(Nm^2 G)^3}{V} \right)^{1/3}. \tag{6.122}$$

As it could be anticipated, this temperature is inversely proportional to the volume and is directly proportional to the mass of the system and the gravitation constant because the gas tends to be compressed to a point under self-gravitation but the thermal energy is resistant to this process.

Analogously, we can obtain the critical mass of the system as

$$M_c \equiv m N_c = \left(\frac{V \pi^3 (2 + \ln 2)}{36} \right)^{1/3} \frac{kT}{mG}. \tag{6.123}$$

If the mass of the system is greater than the critical mass for given volume and temperature, then the system collapses.

We can see that finiteness of the volume and the number of particles prevent the collapse for temperatures of the system higher than (6.58) under constant volume and mass or for the mass of the system that is smaller than (6.59) under constant volume and temperature. On the contrary, in the limiting case $N \to \infty$ and $V \to \infty$, the collapse occurs.

The critical point

Let us estimate the critical temperature. Suppose the particles are absolutely hard spheres of volume $v_0 = (\pi/6)a^3$, where a is the diameter, and the system is contained in a vessel-limited region $\Lambda \subset R^\nu, \nu = 3$ so that $x_i \in \Lambda$ for $i = 1, \ldots, N$. Then N particles form the configuration $\Lambda^N \subset (R^\nu)^N$ as

$$\frac{\xi^N}{N!} \exp\left[-\beta U(x_1, \ldots, x_N) \right] dx_1, \ldots, dx_N. \tag{6.124}$$

Configurations excluded from the configuration space are

$$S_a^N = \{(x_1, \ldots, x_N) \in (R^\nu)^N : |x_j - x_j| < a\} \text{ for all pairs } i \neq j. \tag{6.125}$$

Then the space of admitted configuration is $\Lambda^N \backslash S_a^N$ [52]. This means that the system cannot have volume smaller than $V_0 \approx N v_0$. This observation introduces essential modifications in the thermodynamic state of our system.

In order to reveal the character of such modifications we consider the isotherms of the self-gravitating gas.

Equation (6.51) determines the minimum volume after which further compression is accompanied by the collapse, i.e.,

$$V_m = \frac{36}{\pi^3 (2 + \ln 2)} \left(\frac{Nm^2 G}{kT} \right)^3.$$
(6.126)

If the system has volume greater than the latter, then the gas produces positive pressure $\sim 1/V$ (for constant temperature) and is characterized by a spatially homogeneous distribution.

The temperature is critical if the gas cannot collapse for any admissible volumes for temperatures higher than that. Then the critical point is produced by an intersection of the volume V_m and the minimum volume V_0. Solution of the equation $V_m = V_0$ yields

$$kT_c \sim \frac{Nm^2 G}{V_0^{1/3}}.$$
(6.127)

This formula shows that the critical temperature increases with the increase of the initial gravitation energy of the gas.

Nature of the phenomenon

In this section, we shall disclose the phenomenon of the gravothermal catastrophe using simple but clear reasoning. Let the system be a gravitating gas of mass mN that occupies volume V. This gas produces gravitation field

$$\Delta \phi = 4\pi G \rho,$$

where $\rho \simeq mN/V$ and $V \sim R_{max}^3$. In the gauge the field is such that $\phi = 0$ at the boundary of the system, then the energy of a particle in the system is given by

$$U = m\phi = -4\pi m G \rho \frac{R - R_{max}^2}{2}.$$
(6.128)

On the other hand, the thermal energy of a particle is equal to $\sim kT$. If the thermal energy is lower than its maximum gravitation energy (in the center of the system), then the gas has negative pressure and tends to compress to a point. And vice versa, if the particle thermal energy is greater than its maximum gravitation energy (by the module), then the gas has positive pressure and tends to expand to infinity. During compression, the gas thermal energy increases since the total energy of the system is constant, i.e.,

$$NkT + W_{\text{grav}} = \text{const} \Rightarrow Nk\Delta T = -\Delta W_{\text{grav}}, \tag{6.129}$$

where $W_{\text{grav}} < 0$ is the gravitation energy of the gas. This leads to the equilibrium being established that determines the spatial distribution function.

As a result of the aforesaid, the pressure in this state vanishes. The action determines the free energy $F = kTS(\tilde{\xi}_d)$ at the saddle point $\tilde{\xi}_d$. If we suppose $\Delta = \xi\alpha^2$ (as it has been shown in Sec. 6.4), then

$$P = -\frac{\partial F}{\partial V} = -kT\left(\frac{\Delta^2}{r_m^2} - \frac{N}{V}\right) = 0. \tag{6.130}$$

This means that the pressure produced by gravitation (negative) compensates the pressure associated with the thermal motion (positive). The pressure in the spatially homogeneous state is positive.

The point of collapse (the values of the temperature and volume for which the collapse occurs) is determined by the equality of the above-mentioned energies, and so we obtain that

$$\frac{Gm^2N}{kTR_{\text{max}}} \sim 1. \tag{6.131}$$

In the third power, this equation reproduces Eq. (6.51).

If $R_{\text{max}} \to \infty$, then $U_0 \to \infty$, i.e., the thermal energy is always lower than the gravitation energy in the center of the system (in this case, any point may be chosen for the center), hence such self-gravitating gas collapses under any thermodynamic conditions.

The process may be considered from another point of view. In Sec. 6.4 we have shown that a self-gravitating system collapses to infinite multitude of clusters of size D_0 each in the case $N \to \infty$, $V \to \infty$. Apparently, the finite system collapses only provided the initial volume (homogeneous phase) is greater than the size of the collapsed phase. That is, we can obtain the condition (6.51) from the inequality $V \geq D_0^3$ for given temperature.

As it has been said before, such result is associated with the long-range character of gravitation. Each particle interacts with all particles rather than only with the nearest neighbours (unlike the systems with short-range interaction or the lattice gas), and interaction between two particles does not make a major contribution into the energy of the system. Indeed, the major contribution into this energy is produced by the interaction between a particle and a spatially inhomogeneous formation.

Gravitating Bose gas of hard spheres without condensate

In this section, we demonstrate the probability of spatially inhomogeneous distributions existing in a system of interacting Bose particles, obtain the conditions of cluster formation in this system, and study the dynamics of cluster formation in the limiting case of high temperatures.

First of all, we consider Bose gas of hard spheres of diameter a. The inverse operator of interaction is given by [3]

$$U_{rr'}^{-1} = \frac{1}{U_0}\delta_{rr'},$$

where $U_0 \to \infty$. Then, using (5.27), we have

$$S = -\frac{V - V_0}{\lambda^3}g_{5/2}(\xi) + \ln(1 - \xi) + (N + 1)\ln\xi$$

$$+ \frac{1}{\omega}\int_0^{V_0} dV \int_0^\infty d^3p \ln(1 - \xi\exp(-\beta\varepsilon_p)\cos\psi), \quad (6.132)$$

where $d^3p \equiv 4\pi p^2 dp$ is the differential volume in the momentum space, $V_0 \approx 2v_0 N$ is the volume that would be occupied by particles if they are accumulated closely to each other,

$$v_0 = \frac{4}{3}\pi\left(\frac{a}{2}\right)^3$$

is the volume of a single particle. The equation for the saddle point is given by

$$\frac{\partial S}{\partial \psi} = \frac{V_0}{\omega} \int_0^\infty d^3 p \frac{\xi \exp(-\beta \varepsilon_p) \sin \psi}{1 - \xi \exp(-\beta \varepsilon_p) \cos \psi} = 0. \tag{6.133}$$

The solution of this equation is given by

$$\psi = \begin{cases} \pi, & R < R_0 \\ 0, & R > R_0 \end{cases}, \tag{6.134}$$

where $R_0 = \sqrt[3]{\frac{3V_0}{4\pi}}$

The solution (6.134) means that as long as two particles cannot come closer than their diameters, the system cannot be compressed to a volume smaller than V_0.

Now we consider a system of particles interacting by gravitation attraction and hard sphere repulsion. For the Newtonian attraction the inverse operator is known as

$$W_{rr'}^{-1} = -\frac{1}{4\pi G m^2} \Delta_r \delta_{rr'},$$

where G is the gravitation constant, m is the particle mass, and Δ_r is the Laplace operator. Then, using the result of the hard sphere model (6.134), we may write the action as

$$S = \frac{1}{2\beta} \int_{V_0}^V \frac{(\nabla \varphi)^2}{4\pi G m^2} dV + \frac{1}{\omega} \int_{V_0}^V dV \int d^3 p \ln \left[1 - \xi e^\varphi \exp\left(-\beta \frac{p^2}{2m}\right) \right]$$

$$+ \frac{1}{\omega} \int_0^{V_0} dV \int d^3 p \ln \left[1 + \xi \exp\left(-\beta \frac{p^2}{2m}\right) \right]$$

$$+ \ln(1 - \xi) + (N+1)\ln\xi = \int_{V_0}^V dV \left[\frac{(\nabla \varphi)^2}{4 r_m} - \frac{1}{\lambda^3} g_{5/2}(\xi e^\varphi) \right]$$

$$+ \frac{V_0}{\lambda^3} f_{5/2}(\xi) + \ln(1 - \xi) + (N+1)\ln\xi, \tag{6.135}$$

where $r_m = 2\pi G m^2 \beta$, and it should be noted that there appears the special Fermi function [53, 54]

$$f_{5/2}(\xi) = \frac{4}{\sqrt{\pi}} \int_0^\infty dx x^2 \ln\left(1 + \xi e^{-x^2}\right) = \sum_{l=1}^\infty (-1)^{l+1} \frac{\xi^l}{l^{5/2}}.$$

It occurs due to the repulsion of hard spheres that changes the behavior of Bose particles statistically.

We introduce a dimensionless quantity $r = R/r_m$ instead of R, a new variable $\sigma = \exp(\varphi/2)$, and denote $\alpha^2 \equiv r_m^3/\lambda^3$. Given that the expression (6.135) contains a logarithm and $0 < \xi < 1$, the condition imposed on the field σ is $0 < \xi\sigma^2 < 1$.

Now, we consider the case $\xi < 1$ that corresponds to the nonoccurrence of the Bose condensate, i.e., $\langle n_o \rangle/V \to 0$. The action (in spherical coordinates) (6.135) may be written with a new above-mentioned variable, i.e.,

$$
S = 4\pi \int_{r_0}^{\infty} \left[\left(\frac{1}{\sigma} \frac{\partial \sigma}{\partial r} \right)^2 - \alpha^2 g_{5/2}\left(\xi\sigma^2\right) \right] r^2 dr
$$

$$
+ \frac{V_0}{\lambda^3} f_{5/2}\left(\xi\right) + (N+1)\ln\xi. \tag{6.136}
$$

The equation for the saddle point is the Lagrange equation for the function (6.136). Here it is given as

$$
\frac{\partial^2 \sigma}{\partial r^2} - \frac{1}{\sigma}\left(\frac{\partial \sigma}{\partial r}\right)^2 + \frac{1}{2}\alpha^2 \frac{\partial g_{5/2}\left(\xi\sigma^2\right)}{\partial \sigma}\sigma^2 \equiv \frac{\partial^2 \sigma}{\partial r^2} - \frac{1}{\sigma}\left(\frac{\partial \sigma}{\partial r}\right)^2
$$

$$
+ \alpha^2\sigma^2 \sum_{l=1}^{\infty} \frac{\xi^l \sigma^{2l-1}}{l^{3/2}} = 0. \tag{6.137}
$$

This equation has no analytical solution. We consider the limiting case $\xi \to 0$ (Boltzmann gas). Then Eq. (6.137) is reduced to

$$
\frac{\partial^2 \sigma}{\partial r^2} - \frac{1}{\sigma}\left(\frac{\partial \sigma}{\partial r}\right)^2 + \xi\alpha^2\sigma^3 = 0. \tag{6.138}
$$

This equation has a soliton solution [3]

$$
\sigma = \frac{\Delta}{\sqrt{\xi}\alpha} \frac{1}{\cosh \Delta r}, \tag{6.139}
$$

where Δ is an unknown integration constant that will be seen below. Any soliton solution corresponds to a spatially inhomogeneous distribution of particles — a finite-size cluster. The relevant asymptotics for (6.139) is $\sigma^2 = 1$ for $r = d$, where d is the cluster size, and $\sigma \to 0$

as $r \to \infty$. This solution describes the particles involved in the inhomogeneous formation of size d and the nonoccurrence of particles at infinity, since in this case the spatial distribution function is

$$\rho(r) = m\xi \frac{1}{\lambda^3} \sigma^2 \qquad (6.140)$$

with the normalization $r_m^3 \int \rho(r) \, d^3r = mN$. We substitute (6.139) in (6.136) bearing in mind that $\lim_{T\to\infty} f_{5/2}(\xi) = \xi$. Thus we have

$$S = 4\pi \int_{r_0}^{d} \left(\Delta^2 - 2\xi\alpha^2\sigma^2\right) r^2 dr + \frac{V_0}{\lambda^3}\xi + (N+1)\ln\xi. \qquad (6.141)$$

Then we carry out integration using the expansion of $1/\cosh x \approx 1 - x^2/2$ in power series of $x \equiv \Delta d \ll 1$, i.e.,

$$S = -(V - V_0)\frac{\Delta^2}{\alpha^2\lambda^3} + \frac{V_0}{\lambda^3}\xi + (N+1)\ln\xi. \qquad (6.142)$$

Here Δ^2 is found from the asymptotics

$$1 = \frac{\Delta^2}{\xi\alpha^2}\left[1 - \Delta^2 d^2\right] \implies \Delta^2 \approx \xi\alpha^2 + \xi^2 d^2\alpha^4.$$

Thus, assuming that $V \gg V_0$, we have

$$S = -\frac{V - 2V_0}{\lambda^3}\xi + (N+1)\ln\xi - \frac{V - V_0}{\lambda^3}\xi^2 d^2\alpha^2, \qquad (6.143)$$

where ξ is found from the saddle point equation $\partial S/\partial\xi = 0$ assuming that $\lambda^3/V \gg \lambda^6/V^2$, $\xi_0 \gg \xi_G$, (ξ_0, and ξ_G are activities of the ideal (5.28) and gravitating Bose gases, respectively, ξ_0^{sph} and ξ_G^{sph} are similar activities with the correction on the particle volume V_0), i.e.,

$$\xi = \frac{\lambda^3(N+1)}{V - V_0} - \frac{2d^2\alpha^2\lambda^6(N+1)^2}{(V - V_0)^2} \approx \frac{\lambda^3(N+1)}{V} + \frac{\lambda^3(N+1)V_0^2}{V^2}$$

$$- \left(\frac{2d^2\alpha^2\lambda^6(N+1)^2}{V^2} + \frac{4d^2\alpha^2\lambda^6(N+1)^2 V_0}{V^3}\right)$$

$$= \xi_0 + \xi_0^{\text{sph}} + \xi_G + \xi_G^{\text{sph}}. \qquad (6.144)$$

Then integrating (6.143) at the saddle point (6.144) in accordance with this formula, we obtain the partition function

$$
Z_N = Z_N^0 \times \exp\left[\frac{V}{\lambda^3}\left(\xi_0^{\text{sph}} + \xi_G + \xi_G^{\text{sph}}\right) - \frac{2V_0}{\lambda^3}\xi - (N+1)\right.
$$

$$
\times \ln\left(1 + \frac{\xi_0^{\text{sph}} + \xi_G + \xi_G^{\text{sph}}}{\xi_0}\right)\right] \times \exp\left[+\frac{V - V_0}{\lambda^3}\xi^2 d^2 \alpha^2\right],
$$

$$(6.145)$$

where Z_N^0 is the partition function of the ideal gas (5.32). The knowledge of the latter makes it possible to find the free energy of the system, i.e.,

$$
F = F_0 - kT\left[\frac{V}{\lambda^3}\left(\xi_0^{\text{sph}} + \xi_G + \xi_G^{\text{sph}}\right) - \frac{2V_0}{\lambda^3}\xi - (N+1)\right.
$$

$$
\times \ln\left(1 + \frac{\xi_0^{\text{sph}} + \xi_G + \xi_G^{\text{sph}}}{\xi_0}\right)\right] - kT\left[+\frac{V - V_0}{\lambda^3}\xi^2 d^2 \alpha^2\right],
$$

$$(6.146)$$

where F_0 is the free energy of the ideal Bose gas (5.33). Minimizing (6.146) by the cluster size $d = D/r_m$ and disregarding the correction on the volume V_0 in the gravitation part of the activity ξ_G^{sph}, since

$$
\frac{\lambda^6 V_0}{V^3} \ll \frac{\lambda^6}{V^2} \ll \frac{\lambda^3}{V},
$$

we have

$$
\frac{\partial F}{\partial d} = -kT\frac{d\alpha^2\lambda^3 2\,(N+1)^2}{V - V_0}\left[1 - 4\frac{d^2\alpha^2\lambda^3\,(N+1)}{V - V_0}\right] = 0
$$

$$(6.147)$$

and obtain the optimum radius of the cluster to be

$$
d_0^2 = \frac{V - V_0}{4N\lambda^3\alpha^2} = \frac{V}{4Nr_m^3}\left(1 - \frac{V_0}{V}\right)
$$

$$(6.148)$$

or in the dimensionless variables

$$D_0^2 = \frac{1}{4}\frac{VkT}{2\pi Gm^2N}\left(1 - \frac{V_0}{V}\right).\tag{6.149}$$

Now we consider the dynamics of cluster formation. To do this we employ the equation of motion given by

$$\frac{\partial D}{\partial t} = -\chi\frac{\partial F}{\partial D},\tag{6.150}$$

where χ is the coefficient of reciprocal dimension of the diffusion mass flow through the cross-section. Making use of (6.147) and (6.149), we have

$$\frac{\partial D}{\partial t} = \frac{\chi NkT}{2D_0^4}\left(-D^3 + DD_0^2\right).\tag{6.151}$$

We denote $\eta \equiv \frac{\chi NkT}{2D_0^4}$. Then we may rewrite Eq. (6.151) in a more suitable form, i.e.,

$$\dot{D} + \eta D^3 - D_0^2\eta D = 0.\tag{6.152}$$

The solution of this equation under the condition that the initial state of the system is spatially homogeneous and the assumption that $D(0)/D_0 = 1/2$ is given by

$$D^2 = \frac{D_0^2}{1 + 3\exp\left(-2\eta D_0^2 t\right)}.\tag{6.153}$$

An analogous result (exponential approach to the equilibrium size) was obtained in paper [162].

Now let us consider the asymptotics of the solution (6.153) for stability. For this, we consider a small deviation from (6.153)

$$X(t) = D(t) - D^o(t),\tag{6.154}$$

where

$$D^0(t) = \left\{\begin{array}{ll} 0 & \text{at} \quad t \to -\infty \\ D_0 & \text{at} \quad t \to +\infty \end{array}\right\}.$$

$D^0(t) = 0$ means that the state of the system is spatially homogeneous, $D^0(t) = D_0$ means that the state is spatially inhomogeneous with the cluster of equilibrium size D_0. We denote $B \equiv (\eta D_0^2 D - \eta D^3)$. Expanding $B(D)$ in power series of $X(t)$ and neglecting powers higher than the first, we obtain an equation for the small deviation, i.e.,

$$\dot{X} = \frac{dB}{dD}\Big|_{D=D^0} X = \left(D_0^2 \eta - 3\eta (D^0)^2\right) X. \qquad (6.155)$$

The solution of this equation is given by

$$X(t) \sim \begin{cases} \exp(D_0^2 \eta t) & \text{at} \quad D^0 = 0 \\ \exp(-2D_0^2 \eta t) & \text{at} \quad D^0 = D_0 \end{cases}. \qquad (6.156)$$

Suppose the initial state of the system is spatially homogeneous ($D = 0$). It is not unstable because the small deviation (6.154) increases exponentially $\sim \exp(kt)$ (where k is the Liapunov exponent in (6.156)). Some fluctuation of density that may occur in the system produces the gradient of the gravitation potential. Then the latter produces spatial inhomogeneity — a cluster with the size approaching the equilibrium value (6.149) asymptotically. This spatially inhomogeneous state with the cluster of size $D = D_0$ is stable because the small deviation exponentially decreases, $\sim \exp(-kt)$.

Bose condensate

Suppose we have two models of the Bose condensate. The first one suggests that particles interact by short-range attraction forces. Such interaction is described by the scattering length $a < 0$ [163]. The other condensate consists of hard spheres with the diameters $a_{\text{sph}} > 0$ interacting through the long-range gravitation forces proportional to $1/R$. The gravitation interaction cannot be described by the scattering length because the change of the S-wave phase under scattering by the potential of effective radius r_0 is given by the expression

$$r_0 = -\frac{1}{a} + \frac{1}{2}k^2 r_0 + \cdots$$

while for the Newton potential $r_0 \to \infty$.

Condensates with negative scattering lengths have been studied both in experimental [27–29] and theoretical [32, 58, 164–167] works.

It has been proved that such systems become unstable with collapse if the number of atoms attains the critical number N_c. Now we compare the properties of the Bose condensate of particles with negative scattering lengths and with long-range attraction to the instability towards collapse.

It should be noticed that the Bose condensate is a continuous wave of matter — a coherent state [168], and hence it is described by some wave function that is the product of one-particle functions in the first approximation [58]. On the other hand, the apparatus of statistical mechanics is based on the postulate of random phases [53] that is not valid in the coherent state. Thus we cannot use the above-mentioned method for the investigation of spatial inhomogeneity in the condensed phase. Hence, in order to study such model systems we apply the method based on the Gross–Pitaevsky equation [30, 31, 163] (we are considering the spherical-symmetry problem only). We have

$$i\hbar\frac{\partial\psi}{\partial t} = -\frac{\hbar^2}{2m}\left(\frac{\partial^2\psi}{\partial R^2} + \frac{2}{R}\frac{\partial\psi}{\partial R}\right) + \frac{4\pi\hbar^2|a|}{m}N\,|\psi|^2\,\psi + V\psi,$$

$$\tag{6.157}$$

$$i\hbar\frac{\partial\varphi}{\partial t} = -\frac{\hbar^2}{2m}\left(\frac{\partial^2\varphi}{\partial R^2} + \frac{2}{R}\frac{\partial\varphi}{\partial R}\right) + \frac{4\pi\hbar^2 a_{\text{sph}}}{m}N\,|\varphi|^2\,\varphi + U\varphi.$$

$$\tag{6.158}$$

Here ψ, φ are the wave functions of the condensates with densities $\rho(R) = mN\,|\psi|^2$ or $\rho(R) = mN\,|\varphi|^2$; m is the mass of a particle, and V is the energy of a particle in the external field (harmonic potential of the trap), i.e.,

$$V = m\omega^2 R^2/2. \tag{6.159}$$

U is the energy of a particle in the gravitation field of the condensate mass distributed by the law $\rho(R)$), i.e.,

$$U = -mG\int\frac{\rho 4\pi R^2 dR}{R}. \tag{6.160}$$

This field produces a trap V due to the property of the long-range action.

Equation (6.157) is a nonlinear differential equation of the second order with variable coefficients; Eq. (6.158) is a nonlinear integral-differential equation of the second order with variable coefficients, Hence, we have to solve them numerically. These equations have soliton solutions under certain conditions that will be found in what follows. Similarly to the previous section, the availability of such solutions implies the occurrence of the spatial inhomogeneity of the system, i.e., a cluster. The stability of the soliton solution of Eq. (6.157) was investigated numerically in [32]. We shall study and compare some general properties of the stability of the soliton solution of Eqs. (6.157) and (6.158) based on the equation for the energy balance.

We assume that the condensate may be characterized by the mean density

$$\rho = \frac{mN}{V},$$

where

$$V = \frac{4\pi}{3} \left(\frac{L}{2}\right)^3$$

is the volume of the system, L is the spatial region occupied by the condensate. Then the potential energy of the condensate as described by Eq. (6.157) is as follows

$$W = \frac{\hbar^2 N}{2mL^2} - \frac{4\pi |a| \hbar^2}{mV} N^2 + \frac{3}{40} mN\omega^2 L^2, \tag{6.161}$$

where the first addend is the energy of the quantum pressure [58] associated with the uncertainty principle. This energy is resistant to the compression of the gas. The second addend is the energy produced by the pseudo-potential [53]. This energy tends to compress the gas.

The third addend is the energy of the condensate in the external field (6.159). This expression is obtained from $\int_0^{L/2} \rho\varphi dV$, where $\varphi = \omega^2 R^2/2$ is the field potential of the trap, $dV = 4\pi R^2 dR$ is the differential volume.

We see that this Bose condensate is unstable. The instability towards collapse occurs due to tunnelling through the barrier of particle attraction and quantum pressure. Let us estimate the length and the height of the barrier. There is no need to find the exact values because the expression (6.161) is approximate and its derivation is an equation of the fifth order!!!. The length of the barrier is given by

$$l \sim \sqrt{\frac{\hbar}{m\omega}} - |a| N. \tag{6.162}$$

The height of the barrier is given by

$$W_m \sim \frac{\hbar^2}{|a|^2 Nm}. \tag{6.163}$$

As follows from formula, (6.163), the barrier vanishes when the number of particles is greater than the critical number,

$$N_c \sim \frac{\sqrt{\hbar/m\omega}}{|a|}. \tag{6.164}$$

Now, we estimate the region occupied by the gravitating Bose condensate and its energy. The potential energy of this condensate described by Eq. (6.158) is given by

$$W = \frac{\hbar^2 N}{2mL^2} + \frac{4\pi a_{\text{sph}} \hbar^2}{mV} N^2 - \frac{9}{10} G (mN)^2 \frac{1}{L}, \tag{6.165}$$

where the first addend is the energy of quantum pressure [58]. The second addend is the energy produced by the pseudo-potential [53]. Given that $a_{\text{sph}} > 0$, this energy tends to expand the gas.

The third addend is the energy in the gravitation field of the condensate mass. This expression is obtained from $(1/2) \int_0^{L/2} \rho\varphi dV$, where $\varphi = -G \int (\rho dV / R)$ is the gravitation potential of the field of the condensate mass.

This energy has a minimum

$$W_0^G \sim -\frac{m^5 N^3 G^2}{\hbar^2} \tag{6.166}$$

at the point

$$L_0^G \sim \frac{\hbar^2}{m^3 NG} \left(1 + \sqrt{1 + \frac{N^2 a_{\mathrm{sph}} Gm^3}{\hbar^2}} \right). \tag{6.167}$$

We see that the gravitating Bose condensate is stable (it cannot collapse) unlike the condensate with the short-range attraction. Such behavior is caused by the long-range action of gravitation and the short-range attraction between particles.

6.5 Models with Attraction and Repulsion

Now we consider a system of particles whose interaction consists of attraction and repulsion. This problem cannot be solved in the general case. Let us reveal the main features of spatially inhomogeneous particle distribution formation in cases when the inverse interaction operator is known. We consider the screened Coulomb repulsion and attraction. Making use of the known form of the inverse operators (5.12), we obtain

$$S = \int dV \left\{ \frac{1}{2r_m} \left[(\nabla \varphi)^2 + \chi^2 \varphi^2 \right] \right.$$
$$\left. + \frac{1}{2r_e} \left[(\nabla \psi)^2 + \lambda^2 \psi^2 \right] - \xi A e^{\varphi} \cos \psi \right\} + (N+1) \ln \xi, \tag{6.168}$$

where $A \equiv \left(2\pi m / \beta \hbar^2 \right)^{3/2}$ as before; χ^{-1} and λ^{-1} are the attraction and repulsion screening radii, respectively; $r_m \equiv 4\pi Q^2 \beta$, $r_e \equiv 4\pi q^2 \beta$; Q^2 and q^2 are interaction constants.

The saddle-point equations are given by

$$\frac{1}{r_m} \left(\Delta \varphi - \chi^2 \varphi \right) + \xi A e^{\varphi} \cos \psi = 0,$$

$$\frac{1}{r_e} \left(\Delta \psi - \lambda^2 \psi \right) - \xi A e^{\varphi} \sin \psi = 0$$

$$\int dV \, \xi A e^{\varphi} \cos \psi = N + 1. \tag{6.169}$$

This set of nonlinear equations determines spatially inhomogeneous field distributions associated with the formation of finite-size clusters. In some cases, these equations can be solved analytically and thus the parameters of such formations can be found.

Let us consider the case when the effective change of the parameters of the system occurs for distances $\lambda^{-1} < r$ and $\chi = 0$. Physically, this corresponds to the long-range attraction and short-range repulsion. Suppose that $\psi \ll 1$. We expand the second equation of the set (6.169) in power series of the "slow" field component ψ to obtain the relation

$$\xi A e^\varphi \frac{\psi^2}{2} = 3\frac{\lambda^2}{r_e} + 3\xi A e^\varphi.$$

Having substituted the latter in (6.168), we may write the effective action as

$$S_{\text{eff}} = 2 \int d\tilde{V} \left\{ \frac{1}{4} (\nabla\varphi)^2 + \xi A r_m^3 e^\varphi + \frac{3}{2} \frac{\lambda^2}{r_e} r_m^3 \right\} + (N+1) \ln \xi,$$

$$(6.170)$$

where the dimensionless length $\tilde{r} = R/r_m$ is introduced and the integration extends over the dimensionless volume $\tilde{V}$. The physical situation described by the effective action (6.170) corresponds to the long-range gravitation attraction and effective repulsion for distances shorter than the interaction radius λ^{-1}.

Let us introduce the notation $\gamma^2 \equiv \xi A r_m^3$ and $\alpha^2 \equiv 3\lambda^2 r_m^3/2r_e$. Then we have the function $\sigma = \exp(\varphi/2)$ in the form

$$S_{\text{eff}} = 2 \int d\tilde{V} \left\{ \left(\frac{1}{\sigma} \frac{d\sigma}{dr} \right)^2 + \gamma^2\sigma^2 + \alpha^2 \right\} + (N+1) \ln \frac{\gamma^2}{A r_m^3}.$$

$$(6.171)$$

This function crucially differs from (5.82): the sign of the second term is opposite and, moreover, it contains an additional term giving rise to the interaction renormalization in the presence of effective repulsion. The extremum condition for (6.171) is realized in the solution of the equation

$$\frac{d^2\sigma}{dr^2} - \frac{1}{\sigma}\left(\frac{d\sigma}{dr}\right)^2 - \gamma^2\sigma^3 = 0 \tag{6.172}$$

with the first integral

$$\left(\frac{1}{\sigma}\nabla\sigma\right)^2 - \gamma^2\sigma^2 = \Delta^2. \tag{6.173}$$

The solution of the latter equation is given by

$$\tilde{\sigma} = \frac{\Delta}{\gamma}\frac{1}{\sinh\Delta\,(r - r')},$$

where r' is the coordinate of the soliton center.

As follows from the distribution function $f(r) = A\xi e^{\varphi}$, this solution describes a spatially inhomogeneous particle distribution. We regard it as a finite-size cluster, with the cluster size to be determined. If a multi-soliton solution is realized, in which soliton centers are dispersed while numbers of particles in the solitons are equal, then $S_{\text{eff}} = nS_{\text{eff}}^0$, where

$$S_{\text{eff}}^0 = \left\{8\pi\int r^2 dr\left[\Delta^2 + \alpha^2 + 2\gamma^2\tilde{\sigma}^2\right] + k\ln\frac{\gamma^2}{Ar_m^3}\right\}. \tag{6.174}$$

Here n is the number of clusters and $k = (N+1)/n$ is the number of particles within a cluster. In a manner similar to the analysis of new phase bubbles formation [24, 160], we write the effective action per cluster, i.e.,

$$S_{\text{eff}}^0 = 8\pi\left\{\frac{1}{3}\tilde{R}^3\left(\Delta^2 + \alpha^2\right) + 2\gamma^2\tilde{R}^2 S_1\right\} + k\ln\frac{\gamma^2}{Ar_m^3}, \tag{6.175}$$

where $\tilde{R}$ is the dimensionless cluster size and

$$S_1 = \int_{2r_0}^{R}\tilde{\sigma}^2 dr = \int_{\sigma_0}^{1}\frac{\sigma\,d\sigma}{\sqrt{\Delta^2 + \gamma^2\sigma^2}}.$$

In the latter expression, the cluster center is assumed to lie at the spherical coordinate system origin. Actually, S_1 describes the cluster surface energy. The above formulas are valid when the transition layer thickness is considerably smaller than the cluster size [33].

For physical reasons, the asymptotic behavior of the solution is the following: $\sigma^2 = 1$ for $r = R$, then $\Delta \sim R^{-1}$ and $\gamma e \sim \Delta$; for $r = 2r_0$, we have $\sigma_0 \simeq \left(4\gamma^2 r_0^2\right)^{-1}$. Thus we obtain $S_1 \simeq - \left(2r_0\gamma^2\right)^{-1}$ and the effective action, in terms of the cluster size, is given by

$$S_{\text{eff}}^0 \simeq 8\pi \left\{ \frac{\tilde{R}^3}{3} \left(\alpha^2 + \frac{1}{\tilde{R}^2} \right) - \frac{\tilde{R}^2}{r_0} \right\} - k \ln \left(A r_m^3 \tilde{R}^2 \right). \qquad (6.176)$$

Minimizing the action with respect to $\tilde{R}$ yields the value of $\tilde{R}$. It is evident that the solution with finite size of the spatially inhomogeneous particle distribution can be realized only for $\alpha^2 R r_0 > 3$. The phase transition occurs for $R = 2r_0$, this corresponds to the condition $2\alpha^2 r_0^2 = 3$. Thus, we obtain the value of the transition temperature as $\theta_c \simeq \pi b^2 Q^2 / r_0$, where $b = (Q/q) \lambda r_0 \gg 1$.

If in this case $\alpha R > 1$, then the effective action reduces to

$$S_{\text{eff}}^0 \simeq \frac{8\pi}{3} \alpha^2 \tilde{R}^3 - k \ln \left(A r_m^3 \tilde{R}^2 \right). \qquad (6.177)$$

Having minimized the action with respect to $\tilde{R}$, we find the cluster size to be $\tilde{R}_0^3 = k/4\pi\alpha^2$, and the action to be given as

$$S_{\text{eff}}^0 = \frac{2}{3}k \left\{ 1 - \frac{6}{\alpha^2 \tilde{R}_0 r_0} - \ln \frac{k}{4\pi\alpha^2} \left(A r_m^3 \right)^{3/2} \right\}. \qquad (6.178)$$

The minimum of this function is realized for the optimum value of the number of particles within a cluster that is determined by the equation

$$k_c = \frac{4\pi\alpha^2}{\left(A r_m^3 \right)^{3/2}} \exp \left(- \frac{24\pi}{\alpha^{4/3} k_c^{1/3} r_0} \right). \qquad (6.179)$$

The critical size of a cluster is $\tilde{R}_c^3 = k_c/4\pi\alpha^2$. Expanding the action (6.178) in power series in the vicinity of the critical value of the number of particles within a cluster, we obtain

$$S^0_{\text{eff}} = \frac{2}{3} k_c \left[1 - \frac{1}{2} \left(\frac{k}{k_c} \right)^2 \right].$$

The probability of finding a cluster of k particles is $P \sim \exp{-S^0_{\text{eff}}}$.

Finally, the free energy of the gas of noninteracting clusters is $F = -\theta \ln Z$. In our case, with regard to the zero modes [178], this reduces to

$$F = n\theta \left\{ S^0_{\text{eff}} - \frac{1}{2} \ln \frac{6}{\pi} S^0_{\text{eff}} \right\}. \tag{6.180}$$

Model with long-range repulsion and short-range attraction

Now let us consider the contrary case when the repulsion range is longer than the attraction range, so that $\lambda = 0$ and $\chi \neq 0$. We assume that $\varphi \ll 1$ and retain only the first term in the action expansion in power series of the field φ. Then, we find from the first equation of the set (6.169) that $\tilde{\varphi} = \left(\xi A r_m / \chi^2 \right) \cos \psi$ and substitute this value in the action (6.168). Thus, we obtain

$$S_{\text{eff}} = \int d\tilde{V} \left\{ \frac{1}{2} (\nabla \psi)^2 - \gamma^2 \cos \psi - \gamma^2 \alpha^2 \cos^2 \psi \right\} + (N+1) \ln \frac{\gamma^2}{A r_e^3}, \tag{6.181}$$

where we have introduced the dimensionless length $\tilde{r} = R/r_e$ and denoted $\gamma^2 \equiv \xi A r_e^3$ and $\alpha^2 \equiv r_m / 2\lambda^2 r_e^3$. Since $\gamma \alpha \ll 1$, we obtain an equation for ψ in the spherically symmetric case, i.e.,

$$\frac{d^2 \psi}{dr^2} + \frac{2}{r} \frac{d\psi}{dr} - \gamma^2 \sin \psi = 0. \tag{6.182}$$

In the general case, the solution of this equation describes a soliton that may be regarded as a spatially inhomogeneous formation. Similarly to (5.83), we may disregard the second term when the dimension of this formation is larger than the transition layer thickness [24, 160]. In this case, the first integral exists, i.e.,

$$\frac{1}{2}\left(\frac{d\psi}{dr}\right)^2 + \gamma^2 \cos\psi = C. \tag{6.183}$$

For $C = \gamma^2$, the solution of Eq. (6.182) is given by

$$\psi = \arctan\exp -\gamma\left(r - r'\right). \tag{6.184}$$

Its asymptotics is $\psi = \pi$ as $r \to 0$, and $\psi = 0$ as $r \to \infty$. Physically this solution describes the formation of a pore in a continuum distribution of particles, i.e., absence of particles within a limited volume of size $d \sim \gamma^{-1}$ that encloses the soliton center r'. In the case of a multi-soliton solution, which corresponds to the formation of a finite number of pores in the system, we may write

$$S_{\text{eff}} = 4\pi n \int_0^d r^2 dr \left\{\left[\frac{4\gamma^2\left(1 + \gamma^2\alpha^2\right)}{\cosh^2\left(yr\right)} - \frac{4\gamma^4\alpha^2}{\cosh^4\left(yr\right)}\right] - \gamma^2\left(1 + \gamma^2\alpha^2\right)\right\}$$

$$- \gamma^2\left(1 + \gamma^2\alpha^2\right)(V - V_d) + (N + 1)\ln\frac{\gamma^2}{Ar_e^3}. \tag{6.185}$$

Here we could use the results of the previous subsection and write the effective action in terms of the pore size. Our purpose, however, is to show the possibility of describing spatially inhomogeneous formations by statistical methods only. In order to obtain the final result we employ the integrals

$$4\pi \int_0^d \frac{r^2 dr}{\cosh^2\left(yr\right)} = \tilde{V}_d \quad \text{and} \quad 4\pi \int_0^d \frac{r^2 dr}{\cosh^4\left(\gamma r\right)} = \frac{5}{3}\tilde{V}_d,$$

where $\tilde{V}_d = (4/3)\pi d^3$. Thus, we obtain the final expression for the action, i.e.,

$$S_{\text{eff}} = 4\gamma^2\left(1 - \frac{2}{3}\gamma^2\alpha^2\right)\tilde{V}_d - \gamma^2\left(1 + \gamma^2\alpha^2\right)\tilde{V} + (N + 1)\ln\frac{\gamma^2}{Ar_e^3}. \tag{6.186}$$

The iteration procedure for γ^2 yields a simplified expression for the action, i.e.,

$$S_{\text{eff}}^0 = -\gamma^2\tilde{V} + (N + 1)\ln\left(\gamma^2/Ar_e^3\right).$$

This corresponds to the approximation that all the pores occupy a volume that is small as compared to the system volume. Thus, we obtain

$$\tilde{\gamma}^2 = (N+1)/\tilde{V}.$$

Substituting this result in the action (6.186), we obtain the final expression for the partition function of the system for the spatially inhomogeneous particle distribution, i.e.,

$$Z_N = Z_N^0 \left\{ 1 + \gamma^2 \alpha^2 - 4\left(1 - \frac{2}{3}\gamma^2\alpha^2\right)\frac{\tilde{V_d}}{\tilde{V}} \right\}^{N+1}. \qquad (6.187)$$

If we set $\gamma\alpha \ll 1$ and $V_d = V_0$ (the particle volume), then we obtain the same result as for the hard spheres model (5.50). Evidently, the phase transition occurs for $\gamma\alpha \to 1$ that corresponds to the temperature $\theta_c \simeq 2\pi\rho Q^2/\lambda^2$. The latter expression may be derived from the condition that the stability of the homogeneous distribution in the system of likely charged particles is violated.

Chapter 7

Cellular Structures
in Condensed Matter

7.1 Cellular Structures and Selection of States

The resulting physical properties systems of interacting particles have been the subject of extensive studies [3, 8, 88]. The phase transitions and the ordering of a many-particle system as well as the properties of the latter are of fundamental importance in physics and for wide practical applications. The most interesting and challenging problem in condensed matter physics is the study of phase transitions with the formation of spatially inhomogeneous distributions of particles, clusters, or cellular structures [3]. These structures were observed in usual colloids [83, 179], in systems of particles introduced in liquid crystals [84, 180, 181], in the case of a ferromagnetic particle in a liquid crystal [85], and even in galactic systems [81]. In some systems, crystal-like ordering such as chain-like structures [182–184] and crystal structures was observed in liquid crystal colloids. As is evident from these examples, the formation of various structures in systems of interacting particles is not essentially dependent on the strength and character of the interaction, it can occur in various physical situations.

This experimental result provides evidence that both determining the conditions and physical motivation of cellular structure formation indicate a very important problem in condensed matter physics. It is worth noting that the appearance of a cellular structure changes dramatically the properties of the condensed matter. Experimentally, it was found that, at the initial stage, a cellular structure originates from a region with the lowest local density of particles. In all cases, the existing interaction leads to a spatially inhomogeneous distribution of particles with the formation of regions free from particles. The cellular structure should realize a minimum of free energy and provide stabilization of regions formed in a pure elastic medium, so that particles are placed at the boundaries of those regions. To describe the conditions for the formation and the properties of these particle structures, one should take into account all aspects of the particle interaction.

For various physical cases, many different specialized explanations of the cellular structure formation have been proposed. Basically, the statistical description of many-particle systems concerns homogeneous states. In order to describe the behavior of a many-particle system with different strengths and features of the interaction, a method is required that would take into account the spatially inhomogeneous distribution of particles [3]. Such a method should consider the formation of a cluster, cellular structures, and structures with lower ordering.

In this chapter, we present a general approach to describe cellular structures in various physical situations for different systems of interacting particles. The theoretical study of a colloidal system is usually focused either on the thermodynamic treatment or on general arguments. The essential point that is ignored by these arguments is that colloidal particles interact via their direct forces and indirect interaction mediated by the solvent phases. Actually, however, they are located within spatially nonuniform distributions of particles and hence interparticle interactions will also influence the physical behavior of the particle distribution and the aggregation process that governs the spatial distribution.

This chapter presents an approach for the thermodynamic description of the cellular structure formation in usual colloids and liquid crystal colloids. It is not difficult to show that the cellular structure formation in such different systems is of similar nature. The formation of different structures depends on the initial concentration of particles and the character of interaction. The system tries to minimize the free energy by breaking it apart, and among many ways to realize it, the system chooses nucleation and further formation of cellular structures. To characterize the aggregation effect is a very difficult task although many attempts have been made to measure fractal dimensional shapes of particle distributions. The method describes the experimental observations of similar systems with various types of interaction between particles. The general approach provides a possibility to estimate the cellular structure formation in various systems under different conditions. The characteristic of a structure (size a cluster or a cell) should depend only on the ratio of the interaction energy to temperature and concentration of particles.

7.2 Thermodynamic of Cellular Structures

The theoretical description of systems of interacting particles is focused on the thermodynamic treatment or on the general arguments. First of all, we consider a system of noninteracting particles. Instead of investigating the nonuniform distribution of particles, we can discuss the probability of formation of voids without particles. To find voids or pores in the distribution of particles is not difficult. In the case of a continuous distribution of particles with concentration $n = N/V$, the size of a pore may be determined as the distance to the nearest particles. We can consider only the formation of spherical voids given that the shapes of voids are not of crucial importance and nonspherical shapes may be considered after a more precise study of various processes of void formation. In the case of a nonuniform distribution of noninteracting particles, the spherical shape of a void is approximated fairly well. To determine the probability of the void formation without particles we employ the Saslaw approach [81] to the gravitational system. The probability of the formation of voids depends on the distribution of nearest particles. For different points in space, the probability $p(r)dr$ to find a particle at a distance between r and $r + dr$ is equal to the probability of nonoccurrence of particles in the area with size r that should be multiplied by the probability to find a particle at distance r. This relation is given by

$$p(r)dr = \left\{ 1 - \int_o^r p(r')dr' \right\} 4\pi n r^2 dr. \qquad (7.1)$$

The latter equation is equivalent to the one given by

$$\frac{d}{dr}\left\{\frac{p(r)}{4\pi nr^2}\right\} = -p(r) = -\left\{\frac{p(r)}{4\pi nr^2}\right\}4\pi nr^2, \tag{7.2}$$

that has a solution

$$p(r) = 4\pi nr^2 \exp\left\{-\frac{4\pi nr^2}{3}\right\}. \tag{7.3}$$

The probability $p(r)dr$ to find a particle at the distance between r and $r + dr$ is equal to the probability of nonoccurrence of particles in the volume V multiplied by the probability ndV to find a particle within the volume between V and $V + dV$, i.e., $p(r)dr = nPdV$. From this relation, we obtain the probability of finding a void without particles with volume V as given by

$$P(V) = \exp(-nV) \equiv \exp(-N). \tag{7.4}$$

The latter expression has the form of Poisson distribution. It is obvious that in a system of noninteracting particles we can obtain only a nonuniform distribution of particles, but there exists a nonzero probability to find a void without particles with volume V.

According to the fundamental principles of thermodynamics we can find an analytical result for the probability to find voids without particles by random selection in volume V in the case of a system of interacting particles. In this case we can employ the standard form of the grand canonical ensemble of N particles in volume V at temperature T [53] given by

$$W(N) = \exp\{\beta\mu N - \beta F\}, \tag{7.5}$$

where $\beta = 1/kT$ is the reciprocal temperature, μ is the chemical potential, and $F(N, V, T)$ is the free energy that may be calculated by the canonical assemble. With this probability, we can find the averages of various physical quantities that depend on the number of particles. As an example, we employ the function z^N [81] where z is an arbitrary variable. For the mean value of this function we obtain

$$\langle z^N \rangle = \sum_N \exp(N \ln z) W(N) = G^{-1} \sum_N \exp\left\{(\beta\mu + \ln z)N - \beta F\right\}$$

$$\equiv \exp\left\{\Psi(\beta\mu + \ln z) - \Psi(\beta\mu)\right\}. \tag{7.6}$$

The latter expression follows from the well-known relation between the partition function of the grand canonical ensemble G and the thermodynamic function, i.e.,

$$lnG(\mu, V, T) = \frac{PV}{kT} \equiv \Psi(\mu, V, T). \tag{7.7}$$

This is the thermodynamic definition of the average values of quantities that depend on the number of particles. We may also propose the statistical definition of the mean values of similar quantities that follows from the probability $P(N)$ to find N particles in volume V, i.e.,

$$\langle z^N \rangle = \sum_N z^N P(N),$$

that should be equivalent to the thermodynamic expression

$$\sum_N z^N P(N) = \exp\left\{\Psi\left[z(\exp\beta\mu)\right] - \Psi\left[\exp(\beta\mu)\right]\right\}$$

$$\equiv \exp\left\{-\Psi\left[\exp(\beta\mu)\right]\right\}\exp\left\{\Psi\left[z\exp(\beta\mu)\right]\right\}. \tag{7.8}$$

Comparing equal number of terms of the power series in z on both sides of the equation thus obtained, we get

$$P(N) = \exp(-\Psi)\frac{\exp(N\beta\mu)}{N!}(\exp\Psi)_0^N, \tag{7.9}$$

where

$$(\exp\Psi)_0^N = \left\{\left(\frac{d}{d(z\exp(\beta\mu))}\right)^N \exp\left[\Psi(z\exp(\beta\mu))\right]\right\} \tag{7.10}$$

may be calculated for $z = 0$. Now, we can calculate the probability to find an empty void in the distribution of interacting particles.

This probability may be written as

$$P(0) = \exp\left(-\frac{PV}{kT}\right). \tag{7.11}$$

If the equation of state is known, then in the general case, we can estimate the probability of voids existing without particles in the system of interacting particles. This relation shows that the existence of a void without particles depends on the equation of state for particles that fill these voids. It is given by a very good representative formula. Considering that the Laplace pressure of a void is $P = 2\sigma/R$, where R is the radius of a void, we obtain the probability in the well-known form

$$P(0) = \exp\left(-\frac{8\pi\sigma R^2}{3kT}\right). \tag{7.12}$$

It is similar to the probability of the formation of a bubble of a new phase under the first-order phase transition. Thus we have a good argument to correct the description of the void formation. We should note that the formation of a cellular structure depends only on the thermodynamic properties of a system and can occur in any system of interacting particles and does not depend on the nature of interaction. Formation of a void produces the surface tension between the void and particles. Formation of a surface void should lead to the decrease of free energy of the system. This formula will be reconstructed by another approach in the next sections of this chapter.

As we have already mentioned, if the equation of state is known, then we can estimate the probability of the existence of voids without particles for various systems of interacting particles. First of all, we consider the thermodynamic properties of a system with weak interaction between particles. In the general case, we can write the equation of state as

$$\frac{PV}{kT} = N(1 - nb_2), \tag{7.13}$$

where b_2 is the second virial coefficient. This is the Van der Waals equation of state where the second part represents the interaction in the system and may be obtained for various interesting situations.

For this system, the probability of a void in the distribution of interacting particles may be written as

$$P(0) = \exp\left\{-\frac{PV}{kT}\right\} \equiv \exp\{-nV(1 - nb_2)\}. \tag{7.14}$$

In the case with no interaction in the system, this formula reduces to the previous result of the Poisson distribution. The mean value of the size of a void without particles may be estimated by comparing the probability of the decrease e times. From this condition the estimated volume of a void is $V = 1/[n(1 - nb_2)]$. In the case of no interaction in the system, $b_2 = 0$, this formula reduces to the previous result. In most cases of systems of interacting particles the virial coefficient b_2 is negative and the volume of void is the smallest as in the case of noninteracting particles. In the case of a gravitating system of particles, the interaction energy has attractive nature and the virial coefficient is positive. This implies that the void volume becomes greater and thus in the gravitating system the empty area is greater. This result was obtained by the direct computer simulation for a system of equal particles with the gravitation Newtonian interaction. This result was obtained by the direct computer simulation for a system of particles with the gravitation Newtonian interaction (Fig. 7.1).

For the next example, we consider the gas of hard spheres. We introduce the packing factor $\nu \equiv NV_0/V$ where V_0 is the volume of one spherical particle. Then the equation of state of the gas of hard spheres may be written as

$$\frac{PV}{kT} = \frac{V}{V_0}\nu\frac{1 + \nu + \nu^2 - \nu^3}{(1 - \nu)^3}. \tag{7.15}$$

In this case, we compare the left part of this equation to one and thus estimate the volume of a void as

$$V = V_0\frac{(1 - \nu)^3}{\nu(1 + \nu + \nu^2 - \nu^3)}.$$

If $\nu \to 1$, then we obtain $V \to 0$. In the case, of compact packing of hard spheres, it is obvious that voids cannot be formed. If $\nu \to 0$, then $V = V_0(1/\nu)$ and we reproduce the previous well-known result.

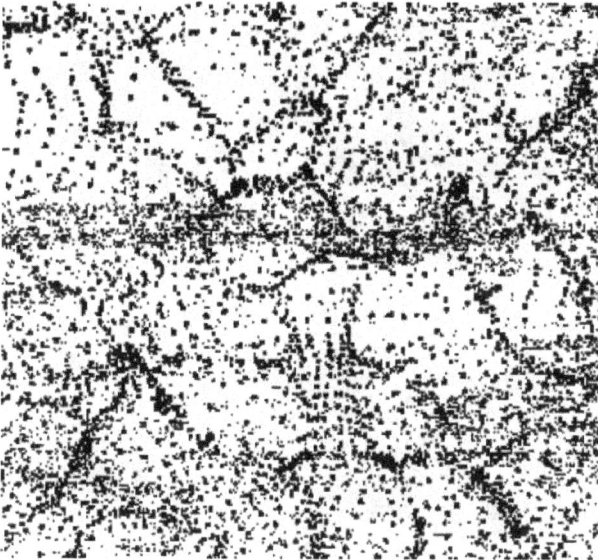

Fig. 7.1.   The result of computer simulation for the system of particles with pure gravitation interaction [81].

7.3   Cellular Structures in Colloids

For the past decade, colloidal systems have been used as model systems in an attempt to understand the phenomenon of the two-dimensional phase transition. In all previous papers dealing with colloidal particles in fluids, attempts have been made to study the phase behavior near phase transitions in solvents with the formation of colloidal voids, soap froths, and clusters [83, 179, 194]. Most papers reported the occurrence of pattern formation by spherical particles trapped at the air–water interface, namely, two-dimensional structures. Depending on the size of a particle, the formation was observed of both reversible and irreversible clustering, that was due to the combination of electrostatic repulsive and long-range van der Waals attractive interactions. The formation of different structures depends on the initial concentration of particles, as one can easily see from the above formulas. The system tries to minimize the energy by breaking it apart, and out of the many ways of doing it, the system chooses the nucleation of voids and further formation of cellular

Fig. 7.2.   The schematic presentation of the cluster or cellular structure formation in the system of interacting particles. The formation of structures depends
on the concentration, interaction energy and temperature.

structures. The difference in probable situations can be seen in the
schematic pictures (Fig. 7.2).

A general description has been proposed related to the formation
of various structures in a system of interacting particles. Basically
a well-known approach is considered for the first-order phase transition. The analysis of this model enables one to predict a rich scenario
of solvent-induced colloidal phase separation with the formation
of clusters and cellular structures. In all cases, phase separation
occurs in two inhomogeneous colloidal phases with different particle

densities [179]. If the interparticle interaction energy is known, then we can study the thermodynamic behavior of an aggregate of such particles and describe the conditions for the formation of a new structure. The character and intensity of the interparticle interaction in the system can produce a spatially inhomogeneous distribution of particles [3]. Phase transition is of the first-order when the external field imposes a surface boundary condition on all particles. Thus the scale is determined of the inhomogeneous distribution of colloidal particles.

Usual colloids

In order to demonstrate the mechanism of phase transition accompanied by the formation of an inhomogeneous distribution, we consider a system of spherical particles. In this case, the free energy of the solution of particles in a usual colloid in a self-consistent field in the many-body approximation may be written as

$$F = F_p + F_s + F_n, \tag{7.16}$$

where

$$F_p = \int U(\mathbf{r} - \mathbf{r}')f(\mathbf{r})f(\mathbf{r}')d\mathbf{r}d\mathbf{r}' + \cdots \tag{7.17}$$

is the free energy in terms of the particle distribution function $f(\mathbf{r})$ and $U(\mathbf{r} - \mathbf{r}')$ presents a sum over all interaction energies. The simplest free energy contains only the pair interaction potential. The entropy part of the free energy is given by the standard expression

$$F_s = kT \int f(\mathbf{r}) \ln f(\mathbf{r}) + [1 - f(\mathbf{r})] \ln[1 - f(\mathbf{r})]d\mathbf{r}. \tag{7.18}$$

The entropy part of the free energy of this type causes the two classical particles not to be able to occupy their individual space positions. Next, the free energy resulting from the coupling between particles and the matter where these are immersed can be modeled as

$$F_n = \int f(\mathbf{r}) \sum_i W(\mathbf{r}_i - \mathbf{r})d\mathbf{r}, \tag{7.19}$$

where $W(r)$ contains microscopic information on the wetting properties of the surface of a particle located at the space point r_i. In the case of liquid crystal colloids this part of the free energy may be presented as the anchoring energy.

The minima of the free energy correspond to the self-consistent field solution for $f(r)$. Each thermodynamic function of state corresponds to a solution that describes some phase of the particle arrangement. If their distribution is inhomogeneous, then the solution helps to find the stable phase associated with the nature of interaction and temperature. If the particle solution is disordered, then by definition the mean value $f(r_i) = c$, where c is the relative particle concentration. The concentration inhomogeneity gives rise to an additional term $f(r) = c \pm \varphi(r)$, where $\varphi(r)$ describes the change of the probability distribution function of particles at different space points. If concentration inhomogeneities are smooth and their scale is much longer than the interparticle distance, then this quantity may be interpreted as the change of particle composition. When considering the continuum description, we can write the free energy increment associated with the inhomogeneous particle distribution by the power series expansion and using the long-wavelength expansion of the concentration, i.e.

$$\varphi(r') = \varphi(r) + \rho_i \partial_i \varphi(r) + \frac{1}{2} \rho_i \rho_j \partial_j \partial_i \varphi(r).$$

Here $\rho = r - r'$ is the distance between particles. In the long-wavelength approximation we can rewrite this part of the free energy in the standard form as

$$\Delta F(\varphi) = \int dr \left\{ \frac{1}{2} l^2 (\nabla \varphi)^2 - \frac{1}{2} \mu^2 \varphi^2 + \frac{1}{4} \lambda \varphi^4 - \varepsilon \varphi \right\}, \qquad (7.20)$$

where $\mu^2 \equiv (V - kT)/(c(1 - c))$, $V = \int U(\rho) d\rho$ and $l^2 = \int U(\rho) \rho^2 d\rho$ is determined through the interaction energy, ρ being the distance between the particles. The coefficient λ is responsible for the nonlinearity of the system that is induced by the many-body interaction in the system of particles. The new coefficient $\varepsilon = N 4\pi R_0^2 W$ represents the energy that is involved in each particle through the wetting effect. Here R_0 is the particle size and W is the energy for anchoring molecules of the elastic medium to the surface of the particle. This coefficient may be introduced provided the wetting effect is similar for each particle. In this case, the general presentation of the free energy should not contain any even terms because the distribution function of macroparticles satisfies the relation $\int f(\mathbf{r})d\mathbf{r} = N$, $\int \varphi(\mathbf{r})d\mathbf{r} = 0$. This expression describes the Landau free energy of a system of particles that are foreign in the elastic medium below the phase-transition temperature of this system. Thus, we can see that the minimum of the free energy realizes a spatially inhomogeneous particle distribution only provided the sign satisfies some relation, and the values of coefficients are determined by the interparticle interaction [66]. In order to find the condition under which the homogeneous particles distribution becomes unstable, we should calculate all the coefficients. The temperature of the phase transition to new states can be determined from the relation $kT_c = c(1 - c)V$ [68].

This is the well-known function that describes the first-order phase transition accompanied by the cluster formation in a system of interacting particles [3, 66]. Thus, the description follows the concentration. The most important contribution to the concentration is associated with the field configuration for which the value of the free energy is minimum, i.e.,

$$\Delta\varphi - \frac{d\Phi}{d\varphi} = 0, \tag{7.21}$$

where the potential is given by

$$\Phi = -\frac{1}{2}\mu^2\varphi^2 + \frac{1}{4}\lambda\varphi^4 - \varepsilon\varphi. \tag{7.22}$$

Substituting the solution of the previous equation in the expression for free energy yields the condition for the formation of a new phase. When the difference of the minima of the effective potential values is greater than the barrier height, then the free energy may be presented in the form

$$\Delta F = 2\pi \int_0^\infty r^2 dr \left\{ \frac{1}{2} \left(\frac{d\varphi}{dr} \right)^2 + \Phi(\varphi) \right\} = -\frac{4\pi}{3} r^3 \varepsilon + 4\pi r^2 \sigma,$$

(7.23)

where σ is the surface energy of the cluster boundary that is equal to the free energy corresponding to the solution of the one-dimensional problem [24, 66, 78], i.e.,

$$\sigma = \int_0^\infty dr \left\{ \frac{3}{2} \left(\frac{d\varphi}{dr} \right)^2 + \Phi(\varphi) \right\} = \int_0^\infty d\varphi \sqrt{2\Phi(\varphi)}. \qquad (7.24)$$

Here the integral should be calculated over the external field. The radius of the cluster can be obtained as $R_c = 2\sigma/\varepsilon$. In our case $\varepsilon = \frac{2\mu\varepsilon}{\lambda^{1/2}}$ and $\sigma = \frac{\mu^3}{3\lambda}$, then $R_c = \frac{\mu^2 l}{3\lambda^{1/2}\varepsilon}$ and the effective value of the free energy variation from the cluster formation is given by

$$\Delta F = \frac{8\pi\sigma R^2}{3}. \qquad (7.25)$$

The probability of the formation of one cluster may be written in the form

$$P(R) = \exp\left\{ -\frac{\Delta F}{kT} \right\} = \exp\left\{ -\frac{8\pi\sigma R^2}{3} \right\}. \qquad (7.26)$$

The characteristic dimensions of the spatially inhomogeneous distribution of particle concentration in the end provide a criterion for the first-order phase transition with cluster formation. The criterion of instability given by this relation may be interpreted as a condition for the formation of a spatially nonuniform distribution for given temperature, which depends on the concentration of particles and on the characteristic length of the new structure.

The void formation may be described in a similar manner by introducing new variables $g(\mathbf{r}) \equiv 1 - f(\mathbf{r})$ and $\phi = 1 - \varphi$ that describe the states of the system from the nonoccurence of particles at different points of the space. In terms of the new variable ϕ, we have an expression for free energy similar to the previous form but with new coefficients, i.e.,

$$\tilde{\mu}^2 = \left(V - \frac{kT}{c(1-c)}\right), \quad V = 4\int U(\rho)d\rho,$$

$\tilde{\lambda} = \lambda$, and $\tilde{\varepsilon} = \varepsilon - 14\int U(\rho)d\rho$ which are determined by the previous coefficients. This renormalization is done to bring an additional part of the free energy ΔF_ϕ rather than to change the nonlinear coefficients. In terms of the new function ϕ, we can obtain the size of a cell as a cluster of voids. In the present approach, the sizes of voids as a new phase of voids without particles may be obtained similarly to the previous case, we thus have $\tilde{R}_c = 2\tilde{\sigma}/\tilde{\varepsilon}$. In our case

$$\tilde{\varepsilon} = \frac{2\tilde{\mu}\varepsilon}{\lambda^{1/4}}, \quad \tilde{\sigma} = \frac{\tilde{\mu}^3}{3\lambda},$$

then

$$\tilde{R}_c = \frac{\widetilde{\mu^2 l^2}}{3\lambda^{1/2}\tilde{\varepsilon}}.$$

This formula suggests that the size of a void is much greater than the size of a cluster that can be formed in this system too. The probability of the formation of a void may be written in a similar form as with new variables from the thermodynamic description of the cellular structure formation. The size of a cell depends on the concentration of particles. This general formula for the probability of formation of voids without particles is similar to the formula that presents the general description of the cluster formation of the new phase. To conclude this section, we have independently estimated the spontaneous formation of loosely bound ordered aggregates of colloidal particles and possible formation of cellular structures due to different nature of interaction. Thus, phase separation in colloidal fluids is directly related to the percolation transition of the wetting solvent phase. The solvent induced phase separation is driven energetically. In the present approach, we can estimate the sizes of voids as the characteristic length of the instability of the nonuniform distribution of particles.

Liquid crystal colloids

Now we employ the results of [77–79, 82–85, 89, 97–133, 146, 147, 154–159, 161–177, 179–193] to briefly present the picture of the cellular structure formation in liquid crystal colloids. Dispersed liquid crystals with macroscopic particles of a foreign substance are a particular feature of such systems. The particles form various structures as observed experimentally [183–185]. It is well-known that the order greatly influences the electro-optical and rheological properties of colloidal dispersions. The system of particles in a liquid crystal changes the liquid crystal state to the soft solid [180, 181] (Fig. 7.3).

The liquid crystal exists in two phases — the isotropic phase, when we observe nonorientation ordering in the long axis of molecules, and the ordered phase, when orientation ordering occurs in the long axis of molecules of the liquid crystal. The particles introduced in such an isotropic phase of a liquid crystal can cause orientation ordering at the expense of the formation of a solvent area. The existence of such deformed areas leads to effective interaction, when these areas overlap [192, 195–197]. Such interactions exist at short distances. In the nematic phase, orientation ordering occurs at long distances. The deformation of the elastic field causes the long-range interaction between particles that are included in the nematic liquid crystal [187–194, 198].

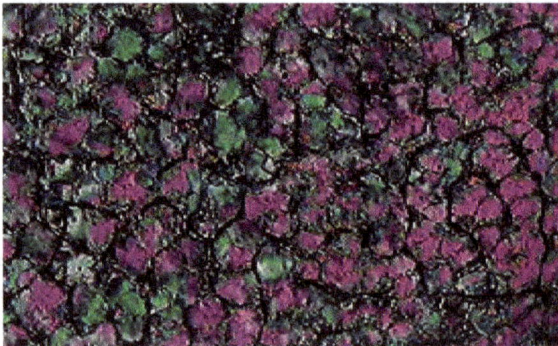

Fig. 7.3. Cellular structure in system: hard spherical particle ZrO_2 in liquid crystal 8CB.

Actually, however, different interactions occur at different distances. At short distances, interactions are induced by the change of the scalar parameter while at long distances the interactions are induced by the deformation in the elastic director field. In the case of low concentrations, interaction between two spherical particles was obtained in [195–197]. For small nanometer-sized particles immersed in the liquid crystal phase, we may assume that each particle changes the order parameter. If the number of particles in the local area increases, then the change of the order parameter increases too. For this process, we cannot take into account the distributional change of the director field. For the case of high concentration of particles introduced into the matter, a more consecutive account of their collective effect is presented in [189, 192]. Interaction can cause new ordering in the system of introduced particles. The effect of ordering is important in the new phase formation in the substance.

In this section, we consider the behavior of a small particle in a liquid crystal and obtain the size characteristic in the formation of a cluster as a function of the concentration and temperature of the medium. The distortions of the order parameter produced by a small particle can lead to the effective interaction between particles and thus motivate the segregation process. We show that this interaction is responsible for the "cellular" and cluster texture in a system of small particles immersed in a liquid crystal. In order to describe such a system, we start from the volume elastic free energy density in the form given by [135]

$$f_b = \frac{1}{2}\Big\{ AQ_{ij}(\mathbf{r})Q_{ij}(\mathbf{R}) + LQ_{ij,k}(\mathbf{r})Q_{ij,k}(\mathbf{r}) \Big\}, \qquad (7.27)$$

where comma indicates both derivation and summation over the repeated indexes. Here A is a positive constant, because we describe the liquid crystal state after phase transition, and $L > 0$ quantifies the cost of creating a distortion in the nematic phase. For the sake of simplicity, we apply the one-constant approximation. Each particle locally changes the order parameter and this fact may be taken into account through the free energy density in the form

$$f_c = \frac{1}{2}W_{ij}f(\mathbf{r})Q_{ij}(\mathbf{r}), \qquad (7.28)$$

where W_{ij} in the general case is the tensor coefficient that determines the interaction between particles and the order parameter, and $f(\mathbf{r})$ is the distribution function of particles in the liquid crystal. The order parameters in the volume may be written in the well-known form

$$Q_{ij}(\mathbf{r}) = S(\mathbf{r}) \left\{ \mathbf{n}_i \mathbf{n}_j - \frac{1}{3} \delta_{ij} \right\}, \tag{7.29}$$

where $\mathbf{n}$ is the director, $S(\mathbf{r})$ presents the order parameter in the volume of the liquid crystal. We can present the free energy for a many-particle system as given by

$$F_b = \frac{L}{2} \int d\mathbf{r} \left\{ Q_{nj,k}(\mathbf{r}) Q_{ij,k}(\mathbf{r}) + \frac{1}{\xi^2} Q_{ij}(\mathbf{r}) Q_{ij}(\mathbf{r}) + w_{ij} f(\mathbf{r}) Q_{ij}(\mathbf{r}) \right\}, \tag{7.30}$$

where the correlation length $\xi^2 = L/|A|$ and $w_{ij} = W_{ij}/L$ is introduced to represent the normalized coupling constant. This free energy describes the behavior of the liquid crystal with particles. The changes caused by individual particles introduced in the medium are developed and produce the average field of deformations of the order parameter. This makes it possible to correctly take into account the collective effect of all particles and to find the self-consistent interaction in the case of high concentration. The minimum of the free energy

$$\frac{\delta}{\delta Q_{ij}(\mathbf{r})} F = 0$$

yields the Euler–Lagrange equation

$$\Delta Q_{ij}(\mathbf{r}) - \frac{1}{\xi^2} Q_{ij}(\mathbf{r}) = w_{ij} f(\mathbf{r}). \tag{7.31}$$

The solution of this equation can be presented in the form

$$Q_{ij}(\mathbf{r}) = w_{ij} \int d\mathbf{r}' f(\mathbf{r}') G(\mathbf{r}, \mathbf{r}'), \tag{7.32}$$

where $G(\mathbf{r}, \mathbf{r}')$ is the Green function of the previous equation. Its form is well known, i.e.,

$$G(\mathbf{r}, \mathbf{r}') = \frac{1}{|\mathbf{r} - \mathbf{r}'|} \exp \left\{ -\frac{|\mathbf{r} - \mathbf{r}'|}{\xi} \right\}. \tag{7.33}$$

This result presents the solution for the distribution of the order parameter by particles with the distribution function $f(\mathbf{r})$ included in the liquid crystal phase.

Now we substitute this expression for $Q_{ij}(\mathbf{q})$ again in free energy and thus obtain the free energy produced by all particles in the liquid crystal as given by

$$F_b = \frac{Lw_{ij}^2}{2\xi^2} \int d\mathbf{r} \int d\mathbf{r}' f(\mathbf{r}) f(\mathbf{r}') G(\mathbf{r}, \mathbf{r}'). \tag{7.34}$$

This free energy is associated with the energy of interaction between areas with particle concentrations $f(\mathbf{r})$ at different space points. The immersed particles change the order parameter as a function of concentration; the changes of the order parameter produce the effective interaction between them.

Having found the elastic free energy of included particles, we can study the thermodynamic behavior of an aggregate of nanoparticles and describe the conditions for the formation of a new structure. The criterion of instability in the homogeneous distribution of particles can be interpreted as a condition for the formation of spatially nonuniform distribution for given temperature that depends on the concentration of particles and on the characteristic length of the new structure. In order to determine this condition, we have to add to the elastic free energy the entropy part that may be written in the standard as

$$F_s = kT \int \left\{ f(\mathbf{r}) \ln f(\mathbf{r}) + [1 - f(\mathbf{r})] \ln[1 - f(\mathbf{r})] d\mathbf{r} \right\}. \tag{7.35}$$

The entropy part of the free energy is responsible for the classical particles that do not occupy similar spatial positions.

The minimum of the two-part free energy $F_b + F_s$ is associated with the self-consistent field solution for the distribution function $f(\mathbf{r})$. This function corresponds to the solution that describes some thermodynamic state of the particle arrangement phase. If their distribution is inhomogeneous, then the solution serves to find the stable phase associated with the nature of interaction and temperature. If the particle solution is disordered, then by definition, the mean value

$f(\mathbf{r}) = c$, where c is the relative particle concentration. The concentration inhomogeneity gives rise to an additional term $f(\mathbf{r}) = c \pm \varphi(\mathbf{r})$, where $\varphi(\mathbf{r})$ is the change of the particle distribution function. If concentration inhomogeneities are smooth and their scale is much longer than the interparticle distance, then this quantity may be interpreted as the change of particle composition. We can write the free energy growth rate associated with the inhomogeneous particle distribution by the power series expansion and using the long-wavelength expansion of the concentration, i.e.,

$$\varphi(\mathbf{r}') = \varphi(\mathbf{r}) + \rho_i \partial_i \varphi(\mathbf{r}) + \frac{1}{2}\rho_i \rho_j \partial_j \partial_i \varphi(\mathbf{r}).$$

Here $\rho = \mathbf{r} - \mathbf{r}'$ is the distance between two different space positions. In this case, we may rewrite the free energy, that depends on the change of the distribution function of particles, in the standard form given by

$$\Delta F(\varphi) = \frac{1}{2} \int d\mathbf{r} \left\{ l^2 \left(\nabla \varphi\right)^2 + \mu^2 \varphi^2 + \right\}, \qquad (7.36)$$

where

$$\mu^2 \equiv \left(\frac{kT}{c(1-c)} + V \right), \quad V = \frac{Lw_{ij}^2}{2\xi^2} \int \rho G(\rho) d\rho,$$

and

$$l^2 = \frac{Lw_{ij}^2}{2\xi^2} \int \rho^3 G(\rho) d\rho.$$

In our case, we can obtain

$$\mu^2 \approx \frac{kT}{c(1-c)} + 2\pi L w_{ij}^2$$

and

$$l^2 \approx +12\pi L w_{ij}^2 \xi^2.$$

This system is always unstable and the length of the first instability is

$$\lambda = \sqrt{l^2/\mu^2} \sim 2\xi \qquad (7.37)$$

for

$$\frac{kT}{c(1-c)} < 2\pi Lw_{ij}^2,$$

ξ is the correlation length. Under the reverse condition, we observe the smallest length of the inhomogeneous particle distribution. In such a case, the Coulomb-like attraction between particles separated by the distance of few nanometers results in the formation of a cluster with strongly interacting particles [189, 192, 195–197]. In the liquid crystal phase, the correlation length is $\xi \sim 1\mu m$ and it is the average size of a cluster of segregated nanoparticles. Cluster formation in a system of nanoparticles is the natural result of the elastic interaction of areas with different particle concentrations. These particles produce the deformation of the elastic matter and this deformation leads to the segregation of small particles. This behavior motivates the long-range repulsive interaction and the short-range attractive interaction through the changes of the order parameter of the liquid crystal. The equilibrium distribution of particles corresponds to their spatially nonuniform distribution. This spatially nonuniform distribution of the introduced particles leads to areas free of particles. The mutual effect of particles and the medium results in the nonuniform distribution of cooperating particles. Thus some kind of a new soft body is formed, whose properties differ from the properties of the medium [180, 181].

Liquid crystal colloids in nematic phase

In recent years, much attention has been paid to the study of the interaction and phase behavior of colloidal particles dispersed in the ordered-phase liquid crystal [78, 193–195, 198–202]. The cellular structure is observed only in the liquid crystal phase while in the isotropic phase, it is not observed. This visualization is based on the macroscopic sizes of immersed particles. To describe the peculiarities of the behavior of a many-particle system in a liquid crystal one should take into account the interaction via the director elastic field. As was shown previously [154, 198], a foreign particle causes the liquid crystal distortion in a region much greater than particle dimensions and thus leads to an effective interaction with other similar particles via the director field deformation. This particle may also be regarded as a particle surrounded by a "solvating shell" provided the interaction between the particle and the liquid crystal molecules is much stronger than the intermolecular interaction responsible for the liquid crystal formation. The solvating formation may be regarded as a particle of size equal to the size of the deformation coating around the particle. Thus, its interaction with another similar formation may be described by the director field deformation [188]. The validity criterion of this treatment of spherical particles may impose the boundary condition on the director for such a formation. In this sense, the interaction of spherical particles is also related to the elastic director field deformation in the liquid crystal in which these particles are dissolved. The problem of interaction between the particles of the nematic liquid crystals has a solution in the self-consistent molecular field approximation when the director field distribution on the surface of an individual particle is determined by the joint effect of all other particles [191]. The effectiveness of such interaction is determined, first of all, by the geometric parameters of the inclusion, by the force of adhesion of the liquid crystal molecules to the surface of such an inclusion, and by the elastic properties of the matter. In the general case, the value and nature of the interaction are determined by the value and nature of the violation of the director distribution symmetry [191]. In the case of small spherical particles, the interaction energy is shown in many papers [25, 154, 198] to have the quadruple form, i.e.,

$$W(\rho) = \left(\frac{4\pi}{15} W R_0^4\right)^2 \frac{3 - 35\cos 0\theta + 35\cos^4 \theta}{\rho^5}. \tag{7.38}$$

Here R_0 is the particle size, ρ is the distance between particles, and θ is the angle between the director and the radius vector of the distance. As is shown in paper [154], the system of particles with interaction of this type is not stable. In what follows, we present some way of solving the problem of void formation in a system of colloidal particles in the liquid crystal, phase. In view of the anisotropic specifics of liquid crystals, we can rewrite the change of the free energy in the form given by

$$F(\varphi) = \frac{1}{2} \int d\mathbf{r} \left\{ \xi_{\parallel} \left(\nabla_{\parallel} \varphi \right)^2 + \xi_{\perp} \left(\nabla_{\perp} \varphi \right)^2 + \mu^2 \varphi^2 \right\}, \qquad (7.39)$$

where

$$\xi_{\parallel} = \frac{1}{2} \int U(\rho) \rho_{\parallel}{}^2 d\mathbf{r}, \quad \text{and} \quad \xi_{\perp} = \frac{1}{2} \int U(\rho) \rho_{\perp}{}^2 d\rho,$$

allows for different probable lengths of the changing distribution of particles along and perpendicular to the director, i.e., $\rho_{\parallel}$, and $\rho_{\perp}$ are the components of the interparticle distance vector. The minimum of the free energy realizes a spatially inhomogeneous particle distribution, provided the sign satisfies some relation and the values of coefficients are determined by the interparticle interaction potential. In order to reveal the condition under which the homogeneous particle distribution becomes unstable, we should calculate all the coefficients [154]. Here, we consider the case when the concentration may be described by the uniform configuration. This concentration wave is formed in accordance with the new equation. We can write the concentration wave in the form of the Fourier transform

$$\varphi(k) = \int d\mathbf{r} \varphi(r) \exp(ikr)$$

of an arbitrary function $\varphi(r)$ defined in a finite volume with periodic boundary conditions. In this case, the relation that determines the inhomogeneity of the particle distribution is given by

$$\frac{kT}{c(1-c)} = -\mu^2 - \xi_{\parallel} k_{\parallel}^2 - \xi_{\perp} k_{\perp}^2. \qquad (7.40)$$

Thus, the homogeneous distribution of particles dispersed in a nematic liquid crystal can be unstable in a limited temperature region where the isotropic phase exists, and a spatially modulated distribution can be generated with the wave vectors $k_{\parallel}$ and $k_{\perp}$.

   In the liquid crystal phase, the estimation of the interaction energy yields the value $V \sim 10^4 kT$ that completely realizes the condition for the formation of a spatially inhomogeneous distribution of particles in this matter [193] (Fig. 7.4). Below we see that the

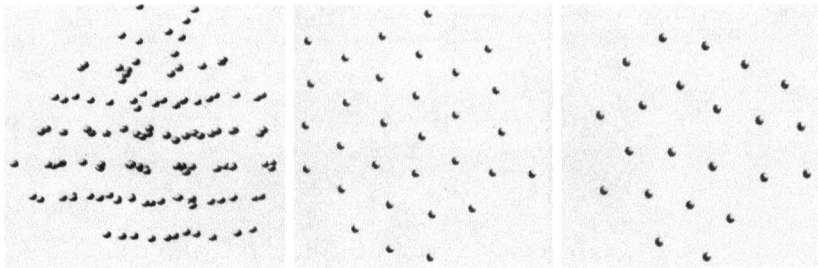

Fig. 7.4.   The result of computer simulation of possible colloidal particle distributions confined to a cubic cell of nematic liquid crystal.

formation of voids in an isotropic phase is impossible while formation of voids in a liquid crystal phase is realized in many cases. The energy of interaction between particles in a liquid crystal is much greater than the energy of interaction in an isotropic phase.

Feronematic colloids

In this section, we briefly present our previous results [188, 191] on another type of liquid crystal colloids, where a cylindrical particle with an intrinsic magnetic moment is immersed into a liquid crystal. This magnetic moment induces interaction of the external magnetic field and the system of particles and can change the orientation of both the particles and the liquid crystal. In principle, however, the rest of the particles without additional moments should effect the interaction potential. The paper [188] examined this case and has shown that deformations from all particles lead to the exponential screening of the pair interaction potential and this fact can explain the cellular structure.

Here, we reproduce this explanation. The interaction potential for a cylindrical particle may be written as [188]

$$U_{pp'} = -\frac{1}{2\pi K} \widehat{A}_l^p \widehat{A}_{l'}^{p'} Q_{l,l'}^+ \frac{\exp(-\xi\rho)}{\rho} \tag{7.41}$$

where operators $\widehat{A}_l^p$ are presented only by the first terms $\widehat{A}_l = \alpha_{lm}(\mathbf{n}_0 \cdot \mathbf{k_m})$ because all others give higher powers in $1/\rho$ [191]. Here

$$Q_{l,l'}^+ = (\mathbf{r_3} \cdot \varkappa_1)(\mathbf{r_1} \cdot \varkappa_{l'}) + (\mathbf{r_2} \cdot \varkappa_l)(\mathbf{r_l} \cdot \varkappa_{l'}),$$

where $\varkappa_l$ represents the orientation of the particle. For the cylinder, the tensor

$$\alpha_{lm} = 2 \oint dsW(s)\nu_l(s)\nu_m(s)$$

has the components [191] $\alpha_{11} = \alpha_{22} = dL\pi W$, $\alpha_{33} = d^2\pi W$, the others $\alpha_{lm} = 0$. Here L is the length of the grain. So $\alpha_{33}/\alpha_{11} = d/L \sim 0.4$ and we should neglect α_{33}. Thus, the interaction energy may be rewritten as

$$U_{\text{cyl}}(R) = -\frac{\alpha_{11}^2 \sin^2\theta \cos^2\theta \exp[-\xi(\theta)\rho]}{4\pi K} \frac{}{\rho}. \qquad (7.42)$$

In the case of equilibrium orientations $\theta = 0, \pi/2$ [203, 204], the screened Coulomb-like interaction does not occur and the higher-order terms remain. It is clear that the macroscopic particle concentration c causes screening of the pair interaction potential with the screening length $\xi^{-1} \approx \sqrt{K/WcS}$ (we imply that W is here the absolute value independent of the sign), S is the area of the particle. The screening occurs both for homeotropic and planar anchoring. The concentration here is included in the inverse screening length ξ only, so that in the limiting case $c \to 0$, we have $\xi = 0$ that brings us back to the unscreened result. In order to describe the experimental results for the dependence of the field-induced birefringence on the strength of the applied field, concentration of the magnetic dopant, and the thickness of the nematic cell paper [188], we introduce the following free-energy density functional given by

$$F = \frac{kT}{v} \int f(\mathbf{r})\ln f(R)dV + \frac{1}{2v^2} \int f(\mathbf{r})f(\mathbf{r}+\rho)U(\rho)d\mathbf{r}d\rho. \qquad (7.43)$$

Here we have to find a condition of stability loss in such a system of attracting particles. It means that the concentration reduces to $f(\mathbf{r}) = c + \delta f(\mathbf{r})$, where c is the ground concentration as before. We perform an expansion in series

$$f(\mathbf{r}+\rho) \approx f(\mathbf{r}) + (\rho\nabla)f(\mathbf{r}) + \frac{1}{2}(\rho\nabla)^2 f(\mathbf{r}).$$

Thus we obtain

$$F - F_0 = \frac{1}{2} \int N \delta f^2(\mathbf{r}) + M(\nabla \delta f)^2, \qquad (7.44)$$

where

$$N = 2kT/v + \frac{1}{v^2} \int_{R_0}^{\infty} U(\rho) d\rho \qquad (7.45)$$

and

$$M = -\frac{1}{2v^2} \int_{r_0}^{\infty} U(\rho) \rho^2 d\rho. \qquad (7.46)$$

Here r_0 is as before the size of the particle. Given that $U < 0$, the phase transition occurs when $N < 0$. In this case, we have $\xi R_0 \ll 1$ and hence we can obtain

$$N \approx \frac{2kT}{f_0 v} - \frac{4\pi e}{\xi^2 v^2} \quad \text{and} \quad M \approx \frac{12\pi e}{\xi^4 v^2}.$$

Below the critical point

$$N \sim \frac{4\pi e}{\xi^2 v^2}.$$

The length of the first instability is

$$l_{\text{inst}} = \sqrt{2M/N} \sim \frac{1}{\xi}. \qquad (7.47)$$

Here

$$\xi^{-1}(\theta) = \sqrt{K/c|a(\theta)|}$$

is the screening length. In the experiment [84] $\xi^{-1} \sim 60\,\mu\text{m}$ as is found above.

For the concentration $c \sim 10^{10}\,\text{cm}^{-3}$, the average distance between particles is $\langle l \rangle \sim 5\,\mu\text{m}$, so that $\xi^{-1} \gg \langle l \rangle$. There are about 1000 particles in volume ξ^{-3} and they induce screening indeed. As we have discussed before, $l_{\text{inst}} \sim 50\,\mu\text{m}$ that is in good agreement with the experimental size of the "cells" [84]. In the experiments by Chen and Amler [84] with the cylindrical particles, the concentration is $c \approx 10^{10}\,\text{cm}^{-3}$, $S \approx RL$. The radius of the grain is $R \approx 0.05\,\mu\text{m}$, the length $L \approx 0.5\,\mu\text{m}$. The elastic constant $K \sim 10^{-7}\,\text{dyn}$, the anchoring energy $W \sim 10^{-3}$ dyn/cm, and we can find $\lambda_{res} \approx 60\,\mu$m. Both concentration and anchoring can be changed so that the resonance range is $\lambda_{res} \sim 1 - 100\,\mu$m.

The cell formation in a system of colloidal particles in a liquid crystal can be responsible for the Janossy effect that consists of the decrease by two orders of the critical field of the Fredericks transition under laser illumination in a liquid crystal with J-aggregates as compared to a pure liquid crystal [205]. J-aggregates in a liquid crystal are oriented with respect to the minimum free energy. This orientation can be neither parallel or perpendicular to the director, being just an equilibrium orientation of individual aggregates. Under laser illumination, each aggregate changes its intrinsic orientation and can cause deformation of the director elastic field that gives rise to the interaction between aggregates as we have described above. This energy of interaction is shown for the formation of cellular structures in a system of aggregates. The formation of a cellular structure in a system of aggregates can be caused by the change of the orientation of the director that may be regarded as the Fredericks transition.

7.4 Geometry of the Distribution of Interacting Particles

Any physical theory is based on the postulated geometric properties of space. The problem of geometry as a whole is equivalent to the problem of the behavior of the fields that form the space [206–210]. We think that in the problem formulated in such a way, the geometric aspect is important not only for the description of the Universe but also for the study of physical phenomena.

In the last years, interest has been growing in the application of the methods of differential geometry in statistical physics and thermodynamics. The geometric approach is represented by two main ways of investigation. The first approach deals with the metric while the other one considers the contact structure of the thermodynamic phase space. Using the metric tensor we can calculate the scalar curvature for some statistical and thermodynamical models and study the effects due to the relation between metric and physical quantities. Another idea is that the curvature is proportional to the reciprocal free energy of the system. The scalar curvature is divergent near the critical point and, due to its relation to the second moments of fluctuations, the stability of the systems can be measured [211, 212].

The average degree of instability of Hamiltonian dynamics can be given by the curvature-related quantities integrated over the whole mechanical (Riemannian) manifold. This fact establishes a link between the dynamical aspect of a given system, the stability or instability of its trajectories, and some global geometric properties of its associated mechanical manifold [213].

Temperature dependence of abstract geometric observables, e.g., averages of curvature fluctuations, has been studied [214] by the Riemann geometrization of Hamiltonian dynamics, where Lyapunov exponents are related to the average curvature properties of submanifolds of the configuration space [215]. Deformations of submanifolds of thermodynamic equilibrium states introduced by continuous contact maps on a phase-space manifold have been considered via the geometric formulation of thermodynamics [216].

In Riemannian geometric approach to thermodynamics, the theory of fluctuations is also considered [217]. In this approach, for the case of two independent thermodynamic variables, the Riemann curvature is assumed to be inversely proportional to the free energy near the critical point [218]. That leads to a partial differential geometric equation for free energy. Following [218], the solution is a generalized homogeneous function of its arguments and specifying the values of critical exponents results in a full-scale equation of state. Later, this postulate was generalized for the cases with more than two independent thermodynamic variables [219].

In condensed matter physics, attempts have already been made to use the geometric approach for the description of the phase space [211–213, 215, 220–222], to find the criteria of phase transitions [78, 211, 212, 221], to study the formation of the phase boundary [223] and the membrane [224]. The geometric aspect of the physical phenomena occurring in condensed matter may be important for the study of structural phase transitions accompanied by the formation of nonuniform distributions of interacting particles [3, 59]. A nonuniform particle distribution can induce an individual effective space that reflects the character and intensity of the interaction in the system. In the Euclidean space we observe particle distributions determined either by the structure of the crystal in which redistribution of spins, dipoles, atoms, molecules occurs, or by the geometry of the sample with fixed boundaries in which the macroscopic structures are formed. But the effective geometry of the particle distribution can be other than Euclidean. This observation is especially important for soft systems with particle distributions governed only by the nature and intensity of the intrinsic interaction. Even in the crystal structures formed by the interaction between atoms, an effective non-Euclidean geometry can occur and be manifest in the existence of quasicrystals, fullerenes etc [225].

Some approaches have been proposed to use the geometric aspects for the description of physical problems of condensed matter. First of all, it is the usage of the direct analogue with the general theory of relativity. Another approach is based on the calculation of the distribution of interacting particles in the curved space. In biophysics and the theory of liquid crystals, we have solved the problem by obtaining new formations from the minimum of the free energy. Nevertheless, very important questions remain unsolved concerning probable intrinsic reasons that induce the changes of the internal geometry in the system of interacting particles as well as how the nature and magnitude of the interaction between particles change this geometry. The pure gravitational interaction in the system of self-gravitating particles leads to the fractal distribution of particles in the space [8].

Many physical and chemical systems (e.g., membranes, vesicles, type I superconductor films) displaying macroscopic patterns and textures in equilibrium have been analyzed within the framework of competing interactions. The related phenomena of structure formation in coloidal suspensions and superlattice formation in adsorbate films on crystalline substrates have also been studied [226]. In condensed matter physics, the phase transitions with the formation of spatially inhomogeneous distributions of particles, clusters or cellular structures are of great interest. It was shown earlier that structure formation in all systems of interacting particles is of similar physical nature [227] and hence we may describe them geometrically in the same way.

The purpose of this section is to find the effective space of a thermodynamically stable distribution of interacting particles and to give a geometric description of structural phase transitions induced by the nature and intensity of the interaction in the system.

Thus we consider a system of many particles. The free energy in the self-consistent field may be written in the many-body approximation as

$$F = F_p + F_s + F_n. \tag{7.48}$$

Here

$$F_p = \int U(\mathbf{r} - \mathbf{r}')f(\mathbf{r})f(\mathbf{r}')d\mathbf{r}d\mathbf{r}' + \cdots \tag{7.49}$$

is the free energy in terms of the particle distribution function $f(\mathbf{r})$, where $U(\mathbf{r} - \mathbf{r}')$ is the sum of all energies of particle interactions,

$$F_s = kT \int \left(f(\mathbf{r}) \ln f(\mathbf{r}) + [1 - f(\mathbf{r})] \ln[1 - f(\mathbf{r})] \right) d\mathbf{r} \tag{7.50}$$

is the entropy part of the free energy,

$$F_n = \int f(\mathbf{r}) \sum_i W(\mathbf{r}_i - \mathbf{r}) d\mathbf{r} \tag{7.51}$$

is the free energy resulting from coupling between particles and matter where $W(r)$ gives the microscopic information about wetting properties of the surface of a particle located at the space point $\mathbf{r}_i$.

The minimum of the free energy (7.48) corresponds to the self-consistent field solution for $f(\mathbf{r})$. If the distribution of particles in the system is homogeneous at each space point, $\mathbf{r}_i$, then $f(\mathbf{r}_i) = c$ where $c = \text{const}$ is the particle concentration. In the case of inhomogeneous particle distribution $f(\mathbf{r}) = c + \varphi(\mathbf{r})$, where $\varphi(\mathbf{r})$ is the deviation of particle concentration from equilibrium at different space points. In the continuum description (when the deviations of c are smooth on the scale much longer than the distance between particles) we can write $\varphi(\mathbf{r})$ as power series expansion

$$\varphi(\mathbf{r}') = \varphi(\mathbf{r}) + \rho_i \partial_i \varphi(\mathbf{r}) + \frac{1}{2}\rho_i \rho_j \partial_j \partial_i \varphi(\mathbf{r}) + \cdots, \qquad (7.52)$$

where $\rho = \mathbf{r} - \mathbf{r}'$ is the distance between two particles. In the long-wavelength approximation, using the expansion (7.52) and taking into account that $\int f(\mathbf{r})d\mathbf{r} = N$, $\int \varphi(\mathbf{r})d\mathbf{r} = 0$, we may rewrite the free energy (7.48) as

$$\Delta F(\varphi) = \int d\mathbf{r} \left\{ \frac{1}{2}l^2 (\nabla\varphi)^2 - \frac{1}{2}\mu^2 \varphi^2 + \frac{1}{4}\lambda\varphi^4 - \varepsilon\varphi \right\}, \qquad (7.53)$$

where the coefficients

$$\mu^2 = V - \frac{kT}{c(1-c)}, \quad V = \int U(\rho)d\rho, \quad l^2 = \int U(\rho)\rho^2 d\rho \qquad (7.54)$$

are determined in terms of the energy of interaction between the particles in the system.

Coefficient λ is responsible for the nonlinearity of the system induced by the many-body interaction. For example, $\lambda \propto \int U(\rho)U(\rho')d\rho d\rho'$, where ρ is the distance between two individual particles and ρ' is the distance between other individual particles. The nonlinearity may be employed to determine the geometry of the effective spatial distribution of interacting particles.

The last coefficient $\varepsilon = N4\pi R_0^2 W$ represents the energy that includes each particle of size R_0 through the wetting effect.

The most important contribution to the concentration is associated with the field configuration for which the value of the free energy (7.53) is minimum, i.e.,

$$\Delta\varphi - \frac{d\Phi}{d\varphi} = 0, \qquad (7.55)$$

where the potential Φ is given by

$$\Phi = -\frac{1}{2}\mu^2 \varphi^2 + \frac{1}{4}\lambda\varphi^4 - \varepsilon\varphi. \qquad (7.56)$$

When the difference of the effective potential minimum values is greater than the barrier height, then the free energy may be written as

$$\Delta F = 2\pi \int_0^\infty r^2 dr \left\{ \frac{1}{2} \left(\frac{d\varphi}{dr} \right)^2 + \Phi(\varphi) \right\} = -\frac{4\pi}{3} r^3 \varepsilon + 4\pi r^2 \sigma, \quad (7.57)$$

where σ is the surface energy of the cluster boundary that is equal to the free energy associated with the solution of the one-dimensional problem [65, 78], i.e.,

$$\sigma = \int_0^\infty dr \left\{ \frac{3}{2} \left(\frac{d\varphi}{dr} \right)^2 + \Phi(\varphi) \right\} = \int_0^\infty d\varphi \sqrt{2\Phi(\varphi)}. \quad (7.58)$$

Each particle distribution formation in condensed matter induces an individual effective space that does not occur in the Euclidean case. The effective space can be observed through the curvature, torsion, and probable realization of the topologies with arbitrary-order symmetries (including fifth, seventh, etc.). The symmetry and topological properties of the effective space are determined first of all by the specifics of the interaction in the system. It is clear that the problem becomes much more complicated, nevertheless the way to solve it still exists. The idea is associated with the method of constrained global optimization for the Thomson problem of charged particles arrangement over a sphere [228–230]. Having solved this problem, one finds the configuration of the deformed lattice drawn on the surface of a given sphere. Here it is appropriate to mention the system of electrons on the surface of liquid helium. The structure of negative ions in liquid helium after the insertion of electrons in helium was original and unusual. Such particles, being localized inside a spherical void, create a bubble in the liquid helium [231]. The solution of the problem of how electrons are located over the sphere is equivalent to the solution of Thomson problem when N point charges on the surface of a unit conducting sphere interact only through their mutual Coulomb forces and try to find the configuration of charges for which the Coulomb energy is minimized.

The system "helium + charges" is demonstrative in the study of nonlinear effects. Electrons repulse each other hence we have a curved space. The many-body interaction in a system of particles determines the nonlinearity in the free energy. The geometry of the space, in turn, could be determined by the interaction. Thus, the nonlinearity in free energy leads to the inhomogeneous geometry of the space that is non-Euclidean. The minimum of the free energy corresponds to a spatially inhomogeneous particle distribution only for specific signs of the coefficients. Both signs and values of these coefficients are determined by the interparticle interaction [227].

The symmetry order is varied including the fifth-order case. This configuration corresponds to the energy minimum. Obtaining the lattice configuration of particle arrangement over the spherical surface that determines the global energy minimum is in principle similar to the case of particle distribution in the Euclidean space [228, 230]. Different lattice configuration on the spherical surface makes it possible to realize the symmetry order that cannot be observed in the usual Euclidean space [225]. This provides a possibility to realize an arbitrary geometry of interacting particles arrangement determined by the interaction character and intensity without additional restrictions of external conditions. A requirement arises to describe a new geometric background that is determined by the distribution of interacting particles and replaces the Euclidean space. This effective geometry should realize the free-energy minimum with regard for the curvature, torsion, or other geometric characteristics induced by the interaction in the system. Demonstrative examples of twisted spaces in the condensed matter are the helical structure of magnetic substances, cholesteric liquid crystals, and DNA, where the character of the microscopic interaction gives rise to a new topological invariant that in the macroscopic representation induces the effective twisted space with the helix proportional to the interaction parameter. The system of electrons on the surface of helium may be a physical example of such type of changes of the geometry. Each electron deforms the surface because electrons interact not only by the Coulomb repulsion but also by the elastic field of deformations. This interaction is attractive and together with the Coulomb repulsion causes the formation of a spatially inhomogeneous distribution of particles that forms a new surface. It is an effective geometric background. As it is known, in the case of high concentration, electrons form bubbles of a new phase [111]. Such formation may be described in geometric terms. So, the main problem is to find the geometric characteristics of the effective space in terms of the physical parameters of the interaction nature and intensity and to apply them to describe various geometric structures [232]. Geometric characteristics of the space are determined by the minimum of the free energy of the system. The presence of geometric characteristics visualizes probable structures occurring in the particle distributions and, moreover, makes it possible to give an adequate description of physical effects associated with the interconversion thereof.

The most important contribution to the concentration is associated with the field configuration for which the value of the free energy (7.53) is minimum, i.e.,

$$\Delta\varphi - \frac{d\Phi}{d\varphi} = 0, \tag{7.59}$$

where the potential Φ is given by

$$\Phi = -\frac{1}{2}\mu^2\varphi^2 + \frac{1}{4}\lambda\varphi^4 - \varepsilon\varphi. \tag{7.60}$$

When the difference of the minimum effective potential values is greater than the barrier height, then the free energy may be written as

$$\Delta F = 2\pi \int_0^\infty r^2 dr \left\{ \frac{1}{2}\left(\frac{d\varphi}{dr}\right)^2 + \Phi(\varphi) \right\} = -\frac{4\pi}{3}r^3\varepsilon + 4\pi r^2\sigma, \tag{7.61}$$

where σ is the surface energy of the cluster boundary that is equal to the free energy associated with the solution of the one-dimensional problem [65, 78], i.e,

$$\sigma = \int_0^\infty dr \left\{ \frac{3}{2}\left(\frac{d\varphi}{dr}\right)^2 + \Phi(\varphi) \right\} = \int_0^\infty d\varphi \sqrt{2\Phi(\varphi)}. \tag{7.62}$$

If the geometry of particle arrangement is changed because of substance redistribution, which is equivalent, say, to the deformation of the initial lattice, then it is necessary to consider the value of this deformation. By analogy to the theory of gravity, we have to allow for the space deformation in the description of the free energy. In most cases the potential energy $U(\mathbf{r} - \mathbf{r}')$ depends on the relative particle position $\mathbf{r} - \mathbf{r}'$. It is more interesting, however, to consider the dependence of the potential energy on the real positions rather that an arbitrary coordinate system. The uncompensated interaction energy associated with the relative particle arrangement can produce the deformation of the initial lattice that provides the free-energy minimum for these deformations. The interaction between particles is assumed to be so strong that their rearrangement produces lattice deformations compensating for the excessive disproportion. Then the requirement arises that these deformations should be in correspondence with the free-energy minimum calculated for these deformations. In this case, all factors of the phenomenological representation of the free energy should depend on the space point where they change the macroscopic characteristics of the system. Even in the case of linear initial potential there is a nonlinearity in the potential part of the free energy and naturally there is an interaction of the geometric characteristics with the order parameter that can play the role of concentration and it is possible to describe the phase transitions in condensed matter of various types. Most importantly, is the dependence of a space point on the value of the order parameter. The description is more pictorial for a continuously distributed substance because in this case one can find the analogies with and differences from the gravity theory. For metric spaces, the procedure involves the metric tensor. The interparticle interaction in the system results in the redistribution of particles that reflects the geometric nature of the space. In this sense, we see the complete analogy to the gravity theory where the distribution of the fundamental scalar field, which in our case, has the meaning of the particle distribution function, forms the very space. The space curvature can be considered in terms of the free energy in the curvilinear coordinate system, i.e.,

$$F(\varphi, g_{\mu\nu}) = \int \sqrt{-g} dV \left\{ R + \frac{1}{2}\gamma g^{\mu\nu} \nabla_\mu \varphi \nabla_\nu \varphi + V(\varphi) \right\}, \qquad (7.63)$$

where the metric tensor $g_{\mu\nu}$ determines the distance between particles that depends on the behavior of the interaction in the system. The free energy density of the background of the uniform distribution $V(\varphi)$ serves for the "cosmological" constant. The space elasticity γ is completely determined by the interaction in the system. Here, we do not specify the form of the potential $V(\varphi)$. But all coefficients contained in it, determined by the energy of interaction between particles, should depend on the space point and the order parameter in this spatial position [3, 59, 68]. The free energy (7.63) written in the curvilinear coordinate system provides a better description of the real physical situation.

The curvature tensor is completely determined by the particle distribution [233]. And from the Einstein equation

$$R_{\mu\nu} = T_{\mu\nu} - \frac{1}{2}g_{\mu\nu}T$$

we get the relation

$$R_{\mu\nu} = \gamma\nabla_\mu\varphi\nabla_\nu\varphi - g_{\mu\nu}V(\varphi), \qquad (7.64)$$

where $T_{\mu\nu}$ is the energy-momentum tensor of the scalar field φ.

The correlation of the particle distribution in the system with the space in which it occurs may be obtained from the free-energy minimum with respect to the particle distribution, i.e.,

$$\frac{1}{\sqrt{g}}\frac{\partial}{\partial x_\mu}\left\{g^{\mu\nu}\sqrt{g}\frac{\partial\varphi}{\partial x_\nu}\right\} - \frac{dV(\varphi)}{d\varphi} = 0. \qquad (7.65)$$

The first integral of this equation does not present the free energy in complete analogy with the free energy in the general theory of relativity where we have

$$F(\varphi) = \int\sqrt{-g}\,dV\{R + 2V(\varphi)\}. \qquad (7.66)$$

Now the equations for the particle distribution geometry can be obtained by minimizing the free energy (7.66) with respect to the metric, i.e.,

$$\frac{\delta F(\varphi)}{\delta g_{\mu\nu}} = \left(\frac{1}{2}g^{\mu\nu}R - R^{\mu\nu}\right) + \frac{\delta V}{\delta g_{\mu\nu}} = 0. \tag{7.67}$$

In this representation, the scalar curvature R is proportional to

$$\frac{dV(\varphi)}{d\varphi^2} = \mu^2 + \lambda\varphi^2, \tag{7.68}$$

that completely confirms the results in [211, 213] and [214]. If the particle distribution is found, we can investigate the geometry of this distribution.

The picture is even more clear physically in the case of a conformal space when the metric tensor is found by multiplying the Euclidean space metric by a scalar function. In the Euclidean space the distribution of particles produces the effective space that corresponds to the conformal metric tensor. Suppose that the suitable conformal transformation is given by $g_{\mu\nu} = \varphi^2 h_{\mu\nu}$, where $h_{\mu\nu}$ is the metric tensor of the Euclidean space. Then the curvature tensor is described by the expression $R_0 = -6\varphi^{-3}\Delta\varphi$, where Δ is the Laplace operator. The free energy for the conformal space may be written as

$$F(\varphi) = \int \sqrt{-g}\,\{R + 2V(\varphi)\}\,dV \equiv \int \left\{\gamma\frac{\varphi^2 R_0}{12} + V(\varphi)\right\}\sqrt{g}dV, \tag{7.69}$$

where

$$\tilde{R}^{\mu\nu} = \frac{1}{48}h^{\mu\nu}R, \quad \tilde{R} = g_{\mu\nu}R^{\mu\nu} = \frac{\varphi^2 R_0}{12}$$

for the observation that $\varphi^2 = h^{\mu\nu}g_{\mu\nu}$ and $g_{\mu\nu}g^{\mu\nu} = 4$ here. Multiplying by $g_{\mu\nu}$ yields

$$\frac{\varphi^2 R}{6} + \frac{1}{\varphi}\frac{dV(\varphi)}{d\varphi} = 0,$$

which is similar to

$$\gamma\Delta\varphi - \frac{dV(\varphi)}{d\varphi} = 0 \tag{7.70}$$

(see also (7.55)) and finally reduces to the particle distribution function in the usual Euclidean space. When we substitute the solution of Eq. (7.70) in the expression of free energy, we can determine the condition for the formation of a new phase.

In the beginning, the distribution of particles is homogeneous in the Euclidean space while the bubble thus formed is a non-Euclidean object. The flat surface where the particles are placed becomes spherical with even different concentration of particles. The formation of a bubble may be presented as a change of space topology, at the expense of the change of curvature with different particle concentrations on concentric surfaces around the center. If we know the dependence of the radius of a separate concentric sphere on particle concentration, we find the dependence of curvature on concentration. In other words, the distribution of particles in a radial direction can be interpreted as the dependence of the curvature radius on concentration. The solution of the field equation determines the effective space that corresponds to the particle distribution.

Vesicles are often used as model cell membranes. When we study the forms of membranes we consider the simplified free energy in the two-dimensional case, i.e.,

$$F(\varphi, g_{\mu\nu}) = \int \sqrt{-g} dS \left\{ R + \frac{1}{2} \gamma g^{\mu\nu} \nabla_\mu \varphi \nabla_\nu \varphi + V(\varphi) \right\}. \qquad (7.71)$$

The magnitude of the field is a function only of the polar angle θ. Because of the symmetry, the order parameter $\varphi(\theta)$ and the shape function $R(\theta)$ can be expanded in the associated Legendre series [234]. We thus have

$$\varphi(\theta) = \sum_{j=1}^{\infty} \varphi_{2j-1} P^1_{2j-1}(\cos\theta),$$

$$R(\theta) = R_0 \left(1 + \sum_{j=1}^{\infty} r_{2j} P_{2j}(\cos\theta) \right).$$

In the case of equilibrium, the free energy is a minimum with respect to R_0, φ_{2j-1}, r_{2j}. The contributions to the free energy from the curvature and gradients in φ are independent in the region $A = \int \sqrt{-g} dS$. The remaining part of the free energy applied in [234] may be written as

$$F_\varphi = A \left[\frac{1}{\int \sqrt{-g} dS} \int \sqrt{-g} dS \left(\frac{1}{2} \mu^2 \varphi^2 + \frac{1}{4} \lambda \varphi^4 \right) \right]. \tag{7.72}$$

It determines the change of the vesicle shape near the transition. In the disordered phase, the vesicle is assumed to have a spherical shape. Near the transition $j = 1$ modes dominate. The coupling of the shape to the order in the membrane leads to a continuous change in the shape and the sphere turns into cylinder, thus changing the topology of the space. For a given shape $R(\theta)$, minimization of the free energy with respect to $\varphi(\theta)$ is equivalent to solving a nonlinear differential equation for $\varphi(\theta)$. Its solution may be closely approximated by the parameter variation function as in the case of a bubble.

In a simple model for tangent-plane order in the vesicle with spherical topology, it was shown by the mean-field theory how the development of such order leads to the change in the shape of the vesicle [234]. The coupling between the in-plane order and the Gaussian curvature results in the continuous change of the vesicle shape (from spherical to cylindrical) with the increase of the in-plane order degree.

The illustration of the effective curved three-dimensional space could be realized by almost crystal structures called quasicrystals. One can form a perfect icosahedral crystal in the curved three-dimensional surface of S^3, a sphere in four dimensions. The frustration arises when we flatten the sphere to fill the flat space [225]. The frustration of the double-twist vector field in $\mathbf{R}^3$ is relieved on S^3 and thus gives rise to polytope-like models of the blue phases of cholesteric liquid crystals. The twisted vector field on S^3 also serves as a model for molten polymers. Polytopes may be thought of as crystals in the non-Euclidean three-dimensional space. Alternatively, we may consider them to be embedded in a higher-dimensional flat space. Lower-dimensional analogues of polytopes are polyhedrons and polygons that may be regarded as, respectively, two- and one-dimensional non-Euclidean crystals embedded in three-dimensional and two-dimensional flat spaces. Since the non-Euclidean space S^3 represents the surface of a sphere in the Euclidean four-dimensional space, we may do analytic geometry in S^3 using four-dimensional Cartesian coordinates. Following hereinafter [225], we admit that any point in $\mathbf{R}^4$ may be expressed as $\mathbf{u} = u_a \mathbf{e}_a$ where $\mathbf{e}_a$ are four basis vectors in $\mathbf{R}^4$. We can easily calculate the geodesic separation between any two points on S^3. Formally, it is possible to generate a three-dimensional rotation using quaternions that are described by the unit vector in S^3. We have the coordinate system in S^3 in which any point written in the quaternion form corresponds to the three-dimensional rotation. It is useful to label the points in S^3 to describe the polytopes. Certain four-dimensional rotation may be expressed clearly in terms of the familiar three-dimensional concepts.

Ginsburg–Landau model of frustrated icosahedral order provides a phenomenological description of supercooled liquids, glasses and solid phases that may be crystalline or quasicrystalline. To construct a free energy for a liquid we have to identify an order parameter that might be nonuniform in the space. To construct a translation and rotation invariant free energy we have to consider the transformation properties of the order parameter. As in [225], when we combine a spatial translation $\mathbf{r} \rightarrow \mathbf{r} + \mathbf{u}$ and rotation $\mathbf{r} \rightarrow \mathbf{r} + \theta \times \mathbf{r}$, we get the order parameter $\varphi_q(\mathbf{r}) \rightarrow \exp(i\mathbf{q} \cdot \mathbf{u} + i\mathbf{q} \cdot \theta \times \mathbf{r})\varphi_q(\mathbf{r})$. To define an order parameter of this type for the polytope-like ordering we have to find an analogue of the reciprocal lattice vector for the polytope. The projection onto hyperspherical harmonics on S^3 is analogous to the Fourier transformation in $\mathbf{R}^3$. Thereby for the density $\varphi(\mathbf{u})$ on S^3 we have

$$Q_{n,m_1,m_2} = \int d\Omega_{\mathbf{u}} Y_{n,m_1,m_2}(\mathbf{u})\varphi(\mathbf{u}). \qquad (7.73)$$

The index n may be identified with the wave-vector $\mathbf{q}$ through the relationship $|\mathbf{q}|^2 = k^2 n(n+2)$ where $k = \pi/5d$ is the curvature of S^3 and d is the average distance between atoms in the icosahedral domains. In the paper [235], one can find the study of the universal property of curvatures in surface models that display a flat phase and rough phase, the criticality of which is described by the Gaussian model. The Hessian relation gives a clear understanding of the universal curvature jump at roughening transition and facet edges. It also provides an effective way of locating the phase boundaries. In the terms introduced in [235] the universal relation between the surface curvature of the equilibrium, crystal shape, and Gaussian coupling constant yields $k = (\pi kT)/(4\sigma d)$. The total free energy associated with $\varphi_q(\mathbf{r})$ consists of the transition–invariant gradient term so that the transition to the true ground state is likely to be of the first order, and the potential terms that begin at the quadratic order in Q_n. Then [225]

$$\Delta F = \frac{K}{2}\sum_q \left\{ |(\nabla - i\mathbf{q}\times\theta)\varphi|^2 + r\varphi^2 + O(\varphi 4) \right\}$$

$$= \frac{1}{2}\sum_n K_n |DQ_n|^2 + r_n Q_n^2 + \cdots , \tag{7.74}$$

where

$$\Delta F = F - F(c),$$

$$DQ_n = \nabla Q_{n,m_1,m_2} - ik\sum_j \mathbf{e}_j \sum_{m_1',m_2'} (L_j^n)_{m_1 m_2 m_1' m_2'} Q_{n,m_1',m_2'}$$

such that

$$Q_n(\mathbf{r}+\mathbf{u}) = \exp\left\{ i\sum_j L_j^n u_j \right\} Q_n(\mathbf{r}).$$

The spikes in the structure function occur for wave numbers corresponding to the negative coefficient of the quadratic term. The ground state of the free energy (7.74) quite likely corresponds to the polytope-type crystals. The general fourth-order theory is considered that combines the translation and orientation order parameters. It is also shown that quasicrystalline and icosahedral liquid-crystal phases occupy a large portion of the phase diagram [225].

Now, we shall show that a non-Euclidean geometry of particle distributions can be realized even in the absence of nonlinearity, For illustration, we consider a conformal space. We can investigate the distribution of the interacting particles in the Euclidean space solving the appropriate equation. We rewrite the free energy in the cylindrical coordinate system and assume that a spatially modulated structure is realized along the OZ axis of this coordinate system. We have to find how the particle distribution profile varies in the direction perpendicular to OZ. The unknown function φ describes the deformation of the surface where the particles are distributed. The equation for the distribution function is equivalent to the equation for the surface profile in the direction perpendicular to selected axis. In our case, this equation is given by

$$\frac{\partial^2 \varphi}{\partial r^2} + \frac{1}{r}\frac{\partial \varphi}{\partial r} + \frac{1}{r^2}\frac{\partial^2 \varphi}{\partial \theta^2} - \frac{1}{\gamma}\left(k^2 - \mu^2\right)\varphi = 0, \tag{7.75}$$

where r is the radius-vector perpendicular to the axis, θ is the polar angle, k is the wave vector of the periodic modulation along the axis. The exact solution of Eq. (7.75) is given by

$$\varphi(r) = C_1 \sqrt{r}\, I_{(1/2)\sqrt{1+m^2}}\left(\sqrt{\frac{k^2 - \mu^2}{\gamma}}\right)\cos m\theta$$

$$+ C_2 \sqrt{r}\, K_{(1/2)\sqrt{1+m^2}}\left(\sqrt{\frac{k^2 - \mu^2}{\gamma}}\right)\cos m\theta, \tag{7.76}$$

where I_ν, K_ν are the Bessel functions of reduced arguments. It is clear from the solution that the curved surface can realize an arbitrary-order symmetry m of particle arrangement and various values of the parameter $\sqrt{(k^2 - \mu^2)/\gamma}$. The depressions and humps correspond to the probabilities of particle presence or absence. All information about the surface profile deformation that reflects the effective particle arrangement is determined only by the nature and intensity of the interaction in the system by the coefficients μ, γ, etc. This is a probable way of effective space formation in a system of interacting particles. Physically, this fact is equivalent to the existence of effective forces responsible for the elasticity of the space realized by the given particle distribution Fig. 7.5. In this sense our result corresponds to the results of [213] associated with the geometric description of phase transitions.

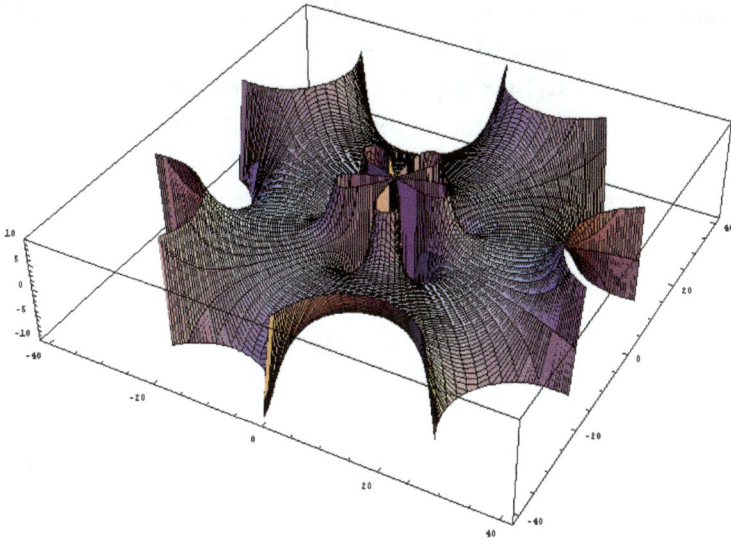

Fig. 7.5.   Distribution of particles on the curved surface.

Thus to conclude, the approach proposed in this section makes it possible to find the effective space by the particle distribution in the system and to describe the visual experimental results in geometric terms. It might be important for the description of quasicrystal states with symmetry of the order such that it cannot be realized in the usual Euclidean space. This requires just allowing for the deformation of the usual crystal lattice or of the surface over which particles with relevant interaction are distributed. Moreover, a possibility arises to describe in geometric terms the structural formation in the macroscopic scale and to investigate their interconversion accompanied by modifying the topology of effective spaces induced by the arrangement of interacting particles. An approach to the description of the physical characteristics of condensed matter in geometric terms is thus offered. Probably in this way, it might be possible to describe the formation of quasicrystals as usual crystals but in the curved three-dimensional space. Moreover, it is possible to describe in geometric terms the phase transition by the creation of a spatially nonuniform distribution of the order parameter. As illustration of the dependence of an effective space curvature on the concentration of interacting particles can be found in the experimental paper [232]. We suppose that the similarity to the general theory of relativity may not only be formal but to have deep origin, as the elasticity of the effective space always may be written in terms of the interpartical interaction. Such effective space depends on the conditions of the system of interacting particles along with their concentration and other average values associated with the macroscopic order parameter that changes the internal geometry of the condensed matter.

Chapter 8

Statistical Description of Nonequilibrium Systems

8.1 Nonequilibrium Gravitating Systems

This part of the book is devoted to a new approach proposed earlier in [1]. It deals with the nonequilibrium statistical operator that is suitable for describing gravitating systems. The equation of state and all thermodynamic characteristics needed for the description are determined by the equations that govern the dominant contribution in the partition function. The approach describes the real inhomogeneous distribution of particles and determines the thermodynamic parameters in a self-gravitating system. The main idea of this section is to provide a detailed description of a self-gravitating system based on the principles of nonequilibrium statistical mechanics and to obtain probable distributions of particles and temperature for fixed number of particles and energy of the self-gravitating system.

Phenomenological thermodynamics is based on the conservation laws for average values of the physical parameters, e.g., the number of particles and energy. Statistical thermodynamics of nonequilibrium systems is also based on the conservation laws, however, for specific dynamical variables rather than for their average values. It presents local conservation laws for the dynamical variables. In order to find the thermodynamic functions in the case of a nonequilibrium system we have to rely on relevant statistical ensembles making allowance for the nonequilibrium states of the systems, under consideration. To describe the nonequilibrium stationary states of the system, we apply the concept of the Gibbs ensemble. In this case, the nonequilibrium ensemble may be determined as a set of systems contained under similar stationary external actions. Such systems show similar character of contact with the thermostat and possess all probable values of thermodynamic parameters corresponding to the given conditions. In the systems that are under similar stationary external conditions, a local equilibrium stationary distribution is formed. If the external condition depends on time, then local equilibrium distributions are not stationary. Exact determination of a local equilibrium ensemble requires the determination of the distribution function or the statistical operator of the system [96]. Finally, we remind that stable states of the systems of equilibrium of classical self-gravitating particles are only metastable because they correspond to the local maxima of the thermodynamic potential.

Under the assumption that nonequilibrium states of a system may be determined by the inhomogeneous distribution energy $H(\mathbf{r})$ and the number of particles (density) $n(\mathbf{r})$, the statistical operator of a local equilibrium distribution for a classical system may be written in the form [96]

$$Q_l = \int D\Gamma \exp\left\{-\int (\beta(\mathbf{r})H(\mathbf{r}) - \eta(\mathbf{r})n(\mathbf{r}))\, d\mathbf{r}\right\}, \qquad (8.1)$$

where the microscopic particle density may be given by the standard expression $n(\mathbf{r}) = \sum_i \delta(\mathbf{r}-\mathbf{r}_i)$. The integration in this formula should be carried out over the whole phase space of the system. It should be mentioned that the Lagrange multipliers $\beta(\mathbf{r})$ and $\eta(\mathbf{r})$ in the case of local equilibrium distribution are functions of the space point. The local equilibrium distribution can be introduced only if the relaxation time in the whole system is greater than the relaxation time in any part of this system.

Once the nonequilibrium statistical operator is determined, we can obtain all thermodynamic parameters of the nonequilibrium system. For this purpose, we determine the thermodynamic relation for the inhomogeneous systems. The variation of the statistical operator by the Lagrange multipliers yields the required thermodynamic relations of the form [96]

$$-\frac{\delta \ln Q_l}{\delta \beta(\mathbf{r})} = \langle H(\mathbf{r}) \rangle_l \quad \text{and} \quad \frac{\delta \ln Q_l}{\delta \eta(\mathbf{r})} = \langle n(\mathbf{r}) \rangle_l \,.$$

These relations make a natural general extension of the well-known relation for the equilibrium systems in the case of inhomogeneous systems. Conservation of the number of particles and energy of the system may be described in terms of natural relations $\int n(\mathbf{r})d\mathbf{r} = N$ and $\int H(\mathbf{r})d\mathbf{r} = E$.

Further statistical description of nonequilibrium systems requires the knowledge of the Hamiltonian of the system. In the general case, the Hamiltonian of a system of interacting particles is given by

$$H = \sum_i \frac{p_i^2}{2m} + \frac{1}{2} \sum_{i,j} W(\mathbf{r}_i \mathbf{r}_j). \tag{8.2}$$

Such presentation of Hamiltonian is valid for spaces of various dimensions. In the three-dimensional case of a real space, the gravitation interaction energy may be written in the well-known form

$$(\mathbf{r}, \mathbf{r}') = \frac{Gm^2}{|\mathbf{r} - \mathbf{r}'|},$$

where G is the gravitation constant and m is the particle mass. In what follows, we write the energy density of a self-gravitating system in the form given by

$$H(\mathbf{r}) = \frac{p^2(\mathbf{r})}{2m} n(\mathbf{r}) + \frac{1}{2} \int W(\mathbf{r}, \mathbf{r}') n(\mathbf{r}) n(\mathbf{r}') d\mathbf{r}'. \tag{8.3}$$

The nonequilibrium statistical operator for the self-gravitating system is given by

$$Q_l = \int D\Gamma \exp\left\{ -\int \left(\beta \frac{p^2}{2m} - \eta \right) n(\mathbf{r}) d\mathbf{r} \right.$$
$$\left. -\frac{1}{2} \int W(\mathbf{r}, \mathbf{r}') n(\mathbf{r}) n(\mathbf{r}') d\mathbf{r} d\mathbf{r}' \right\}. \tag{8.4}$$

Having integrated over the phase space we have

$$D\Gamma = \frac{1}{(2\pi\hbar)^3} \prod_i dr_i dp_i$$

and all the thermodynamic variables $\beta(\mathbf{r})$, $p(\mathbf{r})$, and $\eta(\mathbf{r})$ are functions of the space point. In the next step, we employ the well-known method of field theory. Recently [3, 8], it was shown that the description of this system is exactly equivalent to the field theory of a single scalar field $\varphi(\mathbf{r})$ that contains the same information as the original distribution function, i.e. all information on probable spatial states of the system. In order to perform formal integration in the second part of this presentation, we introduce additional field variables within the context of the theory of Gaussian integrals [7, 12], i.e.,

$$\exp\left\{-\frac{\nu^2}{2}\int \beta\omega(\mathbf{r},\mathbf{r}')n(\mathbf{r})n(\mathbf{r}')d\mathbf{r}d\mathbf{r}'\right\}$$

$$= \int D\varphi \exp\left\{-\frac{\nu^2}{2}\int \beta\omega^{-1}(\mathbf{r},\mathbf{r}')\varphi(\mathbf{r})\varphi(\mathbf{r}')d\mathbf{r}d\mathbf{r}'\right.$$

$$\left. -\nu\int \sqrt{\beta}\varphi(\mathbf{r})n(\mathbf{r})d\mathbf{r}\right\}, \tag{8.5}$$

where

$$D\varphi = \frac{\prod\limits_s d\varphi_s}{\sqrt{\det 2\pi\beta\omega(\mathbf{r},\mathbf{r}')}}$$

and $\omega^{-1}(\mathbf{r},\mathbf{r}')$ is the inverse operator that satisfies the condition $\omega^{-1}(\mathbf{r},\mathbf{r}')\omega(\mathbf{r}',\mathbf{r}'') = \delta(\mathbf{r} - \mathbf{r}'')$. Thus, the interaction energy is represented by the Green function for this operator and $\nu^2 = \pm 1$ depending on the sign of the interaction or the potential energy. The field variable $\varphi(\mathbf{r})$ contains information similar to the original distribution function, i.e., complete information concerning probable spatial states of the system. After the present manipulation, the field variable $\varphi(\mathbf{r})$ contains the same information as the original distribution function, i.e., all information about probable spatial states of the system. The inverse operator $W^{-1}(\mathbf{r},\mathbf{r}')$ of the gravitation interaction in the continuum limiting case should be treated in the operator sense, i.e.,

$$W^{-1}(\mathbf{r}, \mathbf{r}') = -\frac{1}{4\pi G m^2} \Delta_{\mathbf{r}} \delta(\mathbf{r} - \mathbf{r}'), \tag{8.6}$$

where $\Delta_{\mathbf{r}}$ is the Laplace operator in real space. After this manipulation, the statistical operator may be rewritten in the form

$$Q_l = \int D\Gamma D\varphi Q_\varphi \exp\left\{ \int \left[-\beta(\mathbf{r}) \frac{p^2(\mathbf{r})}{2m} + \eta(\mathbf{r}) \right.\right.$$
$$\left.\left. + \sqrt{\beta(\mathbf{r})}\varphi(\mathbf{r}) \right] n(\mathbf{r})d\mathbf{r} \right\}, \tag{8.7}$$

where

$$Q_\varphi = \exp\left\{ -\frac{1}{8\pi m^2 G} \int (\nabla\varphi(\mathbf{r}))^2 \, d\mathbf{r} \right\}. \tag{8.8}$$

The above-mentioned functional integral can be integrated over the phase space. Using the definition of the density, we may rewrite the nonequilibrium statistical operator as

$$Q_l = \int D\varphi D\Gamma \xi(\mathbf{r_i}) \exp\left\{ -\left[\beta(\mathbf{r_i})\frac{p_i^2}{2m} - \sqrt{\beta(\mathbf{r_i})}\varphi(\mathbf{r_i}) \right] \right\} Q(\varphi(\mathbf{r_i})), \tag{8.9}$$

where a new variable $\xi(\mathbf{r}) \equiv \exp\eta(\mathbf{r})$ is introduced, that may be interpreted as the chemical activity, and the definition

$$D\Gamma = \int \frac{1}{(2\pi\hbar)^3 \, N!} \prod_i dr_i dp_i$$

is used. Now it is possible to perform integration over the momentum. The nonequilibrium statistical operator takes the form

$$Q_l = \int D\varphi Q(\varphi(\mathbf{r_i})) \frac{1}{N!} \prod_i \int dr_i \xi(\mathbf{r_i}) \left(\frac{2\pi m}{\hbar^3 \beta(\mathbf{r_i})} \right)^{3/2}$$
$$\times \exp\left[\sqrt{\beta(\mathbf{r_i})}\varphi(\mathbf{r_i}) \right] \tag{8.10}$$

that may be simplified to yield

$$Q_l = \int D\varphi \exp\left\{\int \left[-\frac{(\nabla\varphi(\mathbf{r}))^2}{8\pi m^2 G} + \xi(\mathbf{r})\left(\frac{2\pi m}{\hbar^3 \beta(\mathbf{r})}\right)^{3/2}\right.\right.$$
$$\left.\left. \times \exp\left(\sqrt{\beta(r)}\varphi(\mathbf{r})\right)\right] d\mathbf{r}\right\}. \tag{8.11}$$

In the case of constant temperature β and absolute chemical activity ξ the statistical operator completely reproduces the equilibrium canonical partition function [3, 8]. Following [3, 8], we reduce the nonequilibrium statistical operator to the form

$$Q_l = \int D\varphi \exp\left\{-S\big[\varphi(\mathbf{r}), \xi(\mathbf{r}), \beta(r)\big]\right\} \tag{8.12}$$

where the effective nonequilibrium "local entropy" is given by

$$S = \int \left\{\frac{1}{8\pi m^2 G}(\nabla\varphi(\mathbf{r}))^2 - \xi(\mathbf{r})\left(\frac{2\pi m}{\hbar^2 \beta(\mathbf{r})}\right)^{3/2}\right.$$
$$\left. \times \exp\left[\sqrt{\beta(r)}\varphi(\mathbf{r})\right]\right\} d\mathbf{r}. \tag{8.13}$$

Here we have used the definition of the chemical activity in terms of the chemical potential $\xi(\mathbf{r}) \equiv \exp\eta(\mathbf{r}) \equiv \exp(\mu(\mathbf{r})\beta(\mathbf{r}))$ that depends on the space point. We have already mentioned that in the case of constant temperature β and absolute chemical activity ξ, the statistical operator completely reproduces the equilibrium canonical partition function [3, 8]. The saddle point method can now be employed to find the asymptotic value of the statistical operator Q_l for N tending to ∞; the dominant contribution is given by the states that satisfy the extreme condition for the functional. It is not difficult to see that the saddle point equation presents the thermodynamic relation and may be written in a modified form, i.e., as an equation for the field variable $(\delta S)/(\delta\varphi(\mathbf{r})) = 0$, the normalization condition

$$\frac{\delta S}{\delta(\eta(\mathbf{r}))} = -\int \frac{\delta S}{\delta(\xi(\mathbf{r}))}\xi(\mathbf{r})d\mathbf{r} = N,$$

and the conservation law for the energy of the system

$$-\int \frac{\delta S}{\delta(\beta(\mathbf{r}))} \xi(\mathbf{r}) d\mathbf{r} = E.$$

Solutions of these equations completely determine all the thermo-dynamic parameters and describe the general behavior of the self-gravitating system, whether this distribution of particles is spatially inhomogeneous or not. The above set of equations in principle solves the many-particle problem in the thermodynamic limiting case. The spatially inhomogeneous solution of these equations corresponds to the distribution of interacting particles. Such inhomogeneous behavior is associated with the nature and intensity of interaction. In other words, accumulation of particles in a finite spatial region (formation of a cluster) reflects the spatial distributions of the fields, activity, and temperature. It is very important to note that only in this approach, is it possible to take into account the inhomogeneous distribution of temperature that may depend on the spatial distribution of particles in the system. In other approaches, the dependence of temperature on the space point was introduced through the polytrophic dependence of temperature on the particle density in the equation of state [90]. In the present approach, such dependence follows from necessary thermodynamic conditions and may be determined for various distributions of particles. Now, we derive the saddle-point equation for the extremum of the local entropy function $S(\varphi, \xi, \beta)$. The equation for the field variable $\delta S/\delta\varphi = 0$ yields

$$\frac{1}{r_m} \Delta\varphi(\mathbf{r}) + \xi(\mathbf{r}) \left(\frac{2\pi m}{\hbar^2 \beta(\mathbf{r})}\right)^{3/2} \sqrt{\beta(r)} \exp\left[\sqrt{\beta(r)}\varphi(\mathbf{r})\right] = 0,$$

$$(8.14)$$

where the notation $r_m \equiv 4\pi Gm^2$ is introduced. The normalization condition may be written as

$$\int \xi(\mathbf{r}) \left(\frac{2m}{\hbar^2 \beta(\mathbf{r})}\right)^{3/2} \exp\left[\sqrt{\beta(r)}\varphi(\mathbf{r})\right] d\mathbf{r} = N \qquad (8.15)$$

and the equation for the energy conservation in the system is given by

$$\frac{3}{2} \int \left(\frac{2\pi m}{\hbar^2 \beta(\mathbf{r})} \right)^{3/2} \frac{\xi(\mathbf{r})}{\beta(\mathbf{r})} \left[3 - \sqrt{\beta(r)}\varphi(\mathbf{r}) \right] \exp \left[\sqrt{\beta(r)}\varphi(\mathbf{r}) \right] d\mathbf{r} = E. \tag{8.16}$$

To draw more information on the behavior of a self-gravitating system, we introduce new variables. The normalization condition $\int \rho(\mathbf{r})d\mathbf{r} = N$ yields the definition for the density function, i.e.,

$$\rho(\mathbf{r}) \equiv \left(\frac{2\pi m}{\hbar^2 \beta(\mathbf{r})} \right)^{3/2} \xi(\mathbf{r}) \exp \left[\sqrt{\beta(\mathbf{r})}\varphi(\mathbf{r}) \right], \tag{8.17}$$

that reduces the equation to a simpler form. The equation for the field variable is given by

$$\Delta\varphi(\mathbf{r}) + r_m \sqrt{\beta(\mathbf{r})}\rho(\mathbf{r}) = 0. \tag{8.18}$$

The equation for energy conservation takes the form

$$\frac{1}{2} \int \frac{\rho(\mathbf{r})}{\beta(\mathbf{r})} \left[3 - \sqrt{\beta(r)}\varphi(\mathbf{r}) \right] d\mathbf{r} = E. \tag{8.19}$$

The equation thus obtained cannot be solved in the general case though it is possible to analyze some special cases of the behavior of a self-gravitating system under various external conditions. Hereinafter we write the chemical activity in terms of the chemical potential $\xi(\mathbf{r}) = \exp(\mu(\mathbf{r})\beta(\mathbf{r}))$. Having differentiated the equation for energy conservation over the volume, we obtain an interesting relation for the chemical potential, i.e.,

$$\frac{1}{2}\frac{\rho(\mathbf{r})}{\beta(\mathbf{r})} \left[3 - \sqrt{\beta(r)}\varphi(\mathbf{r}) \right] = \frac{\delta E}{\delta V}\frac{\delta V}{\delta N} = \mu(\mathbf{r})\rho(\mathbf{r}) \tag{8.20}$$

that yields the chemical potential given by

$$\mu(\mathbf{r})\beta(\mathbf{r}) = \frac{3}{2} - \frac{1}{2}\sqrt{\beta(r)}\varphi(\mathbf{r}). \tag{8.21}$$

Within the context of the expression for the density and the definition of the thermal de Broglie wavelength and the gravitation length, i.e.,

$$\Lambda^{-1}(\mathbf{r}) = \left(\frac{2m}{\hbar^2 \beta(\mathbf{r})} \right), \qquad R_g(\mathbf{r}) = 2\pi Gm^2 \beta(\mathbf{r}) \tag{8.22}$$

we can rewrite all the equations and the normalization condition in terms of density and temperature. Thus, we have

$$\Delta \left(\frac{\Lambda^3(\mathbf{r})\rho(\mathbf{r})}{\sqrt{\beta(\mathbf{r})}} \right) + \frac{R_g(\mathbf{r})}{\sqrt{\beta(\mathbf{r})}}\rho(\mathbf{r}) = 0 \qquad (8.23)$$

and the chemical potential reduces to

$$\mu(\mathbf{r})\beta(\mathbf{r}) = \frac{3}{2} - \ln(\Lambda(\mathbf{r})\rho(\mathbf{r})). \qquad (8.24)$$

In this approach, we can also obtain the equation of state for the self-gravitating system by using the thermodynamic relation

$$P = -\frac{1}{\beta}\frac{\delta S}{\delta V}$$

in the case of conservation of energy E. In our case, in the definition of "local entropy" we make use of the relation

$$(\nabla\varphi(\mathbf{r}))^2 = \nabla(\varphi(\mathbf{r})\nabla\varphi(\mathbf{r})) - \varphi(\mathbf{r})\Delta\varphi(\mathbf{r})$$

and after that perform integration over the whole volume. The first part of the integration may be presented as a surface integral where $\varphi(\mathbf{r}) = 0$ on the integration surface. After that, we can present the "local entropy" as

$$S = \int \left[-\frac{1}{8\pi m^2 G}\varphi(\mathbf{r})\Delta\varphi(\mathbf{r}) - \xi(\mathbf{r})\left(\frac{2\pi m}{\hbar^2\beta(\mathbf{r})} \right)^{3/2} \right.$$
$$\left. \times \exp\left(\sqrt{\beta(r)}\varphi(\mathbf{r}) \right) \right] d\mathbf{r} \qquad (8.25)$$

and, using the definition of the particle density, rewrite it in the form

$$S = \int \left[-\rho(\mathbf{r})\ln(\Lambda^3(\mathbf{r})\rho(\mathbf{r})) - \rho(\mathbf{r}) \right] d\mathbf{r}. \qquad (8.26)$$

The local equation of state may be written as

$$P(\mathbf{r})\beta(\mathbf{r}) = \rho(\mathbf{r})(1 - \ln(\Lambda^3(\mathbf{r})\rho(\mathbf{r}))) = \rho(\mathbf{r})\left(\mu(\mathbf{r})\beta(\mathbf{r}) - \frac{1}{2} \right). \quad (8.27)$$

In the classical case $\Lambda^3(\mathbf{r})\rho(\mathbf{r}) \ll 1$ and $P\beta \equiv \rho$, but is logarithmically dependent on the particle density. Only in the case of $\Lambda^3(\mathbf{r})\rho(\mathbf{r}) = 1$, we obtain the usual equation of state for the ideal gas because in this case $\varphi(\mathbf{r}) = 0$ and $P\beta = \rho$ by virtue of nonoccurrence of interaction. For $\mu(\mathbf{r})\beta(\mathbf{r}) = 3/2$, the equation of state thus obtained reproduces the equation of state of the ideal gas. In this case, the energy of the system is equal to $E = (3/2)NkT$ that is in accordance with the results obtained previously. In the next sections, we find the classical distributions of particles for various internal and external conditions.

Homogeneous distribution of particles

First of all, we consider the equilibrium case, all the parameters being independent of the space coordinates. In this case, both energy and total number of particles are fixed and, moreover, the temperature and the chemical potential do not change in space. Thus, the equation for the particle concentration

$$\Delta \left(\frac{\Lambda^3(\mathbf{r})\rho(\mathbf{r})}{\sqrt{\beta(\mathbf{r})}} \right) + \frac{R_g(\mathbf{r})}{\sqrt{\beta(\mathbf{r})}}\rho(\mathbf{r}) = 0 \qquad (8.28)$$

leads to a simple condition $\sqrt{\beta}\rho(\mathbf{r}) = 0$ that can be realized only for $T \to \infty$. The particle distribution in a self-gravitating system can be homogeneous only for very high temperatures.

Another interesting case is when particle density depends on the coordinate while the temperature is fixed. In this case, the equation for the density takes the form

$$\Delta \left(\ln \Lambda^3 \rho(\mathbf{r}) \right) + R_g \rho(\mathbf{r}) = 0 \qquad (8.29)$$

and may be transformed to

$$\Delta \left(\ln \rho(\mathbf{r}) \right) + R_g \rho(\mathbf{r}) = 0. \qquad (8.30)$$

The latter equation has an exact solution

$$\rho(\mathbf{r}) = \frac{2}{R_g r^2},$$

but the normalization condition holds only for the case of a fixed box with size $R = (NGm^2/4kT)$, fixed energy $E = NkT$, and the chemical potential density within the box is given by

$$\mu = kT \left(\frac{3}{2} - \frac{2\Lambda^3}{4kTR_g r^2} \right).$$

As follows from the equation for constant temperature, the homogeneous distribution of particles is unstable. The homogeneous particle distribution $\rho(\mathbf{r}) = \rho + \delta\rho(\mathbf{r})$ yields an equation for density fluctuations, i.e.,

$$\Delta\delta\rho(\mathbf{r}) + R_g\rho\delta\rho(\mathbf{r}) = 0, \tag{8.31}$$

that reproduces the Helmholtz equation. The general solution of the wave equation is the unstable radial distribution $\delta\rho(\mathbf{r}) = \frac{\exp(ikr)}{r}$ with the wave number $k = \sqrt{2\pi Gm^2\beta\rho}$, that implies that the wavelength of the instability is half as long as the Jeans length. It is the statistical length of the instability of particle distribution in the system.

The concept of Jeans gravitation instability is discussed within the framework of nonextensive statistics and is associated with the kinetic theory [118]. A simple analytical formula generalizing the Jeans criterion is derived by assuming that the unperturbed collisionless gas is kinetically described by the class of power-law velocity distributions. It is found that the critical values of the wavelength and mass depend explicitly on the nonextensive parameter. The instability condition is weakened as the system becomes unstable even for wavelengths of the disturbance smaller than the standard Jeans length.

Recent discoveries of extrasolar giant planets, along with the refined models of the compositions of Jupiter and Saturn, suggest a reexamination of the theories of giant planet formation. An alternative to the favoured core accretion hypothesis is examined in [119], the conclusion is that the gravitation instability in the outer solar nebula leads to the formation of giant planets. Three-dimensional hydrodynamic calculations predict the formation of locally isothermal or adiabatic thermodynamics. The gravitation instability appears to be able to form giant planets [120]. Our results can help to explain the data of astrophysical observations in the sense that the different length of the instability in a self-gravitating system is associated with the alternative description of the situation. Thus we can conclude that particle distributions cannot be homogeneous for constant temperatures in the system. Hence, we have to find the true distributions of particles and temperature in the system.

Inhomogeneous distributions of particles and temperature in a gravitating system

In the general case, particle distributions in self-gravitating systems are inhomogeneous. Inhomogeneous distribution of particles gives rise to the long-range gravitation interaction. Now, we consider the nonequilibrium description of a self-gravitating system and take into account probable spatially inhomogeneous distributions of particles and temperature. We introduce a new variable $\psi = \Lambda^3(\mathbf{r})\rho(\mathbf{r})$, then the equation for density is simplified, i.e., we have

$$\Delta\left(\frac{\ln\psi}{\sqrt{\beta(\mathbf{r})}}\right) + \frac{R_g(\mathbf{r})}{\sqrt{\beta(\mathbf{r})}\Lambda^3(\mathbf{r})}\psi = 0. \tag{8.32}$$

The solution of this equation provides a completely nonequilibrium statistical description of a self-gravitating system. General exact solutions of this nonlinear equation are unknown. Further, we propose a way to solve this equation.

First of all, we can find the most general solution of the problem. In the three-dimensional case the action of the Laplace operator may be presented in the form

$$\Delta\left(\frac{\ln\psi}{\sqrt{\beta(\mathbf{r})}}\right) = \frac{1}{\sqrt{\beta}}\left(\frac{d^2}{dr^2} + \frac{2}{r}\frac{d}{dr}\right)\ln\psi$$

$$- \frac{\ln\psi}{\sqrt{\beta^3}}\left[\frac{d^2\beta}{dr^2} + \frac{2}{r}\frac{d\beta}{dr} - \frac{3}{2\beta}\left(\frac{d\beta}{dr}\right)^2\right]$$

$$- \frac{1}{\sqrt{\beta^3}}\frac{d\ln\psi}{dr}\frac{d\beta}{dr}. \tag{8.33}$$

One of the many exact solutions may be obtained if the second term in the right-hand part of this equation is equal to zero. Then the solution for the temperature is given by $\beta = \gamma^3 r^n$ where γ is an unknown constant and it is not difficult to see that for $n = 2$ we obtain only the equation for ψ, i.e.,

$$\frac{d^2 \ln \psi}{dr^2} + \frac{a_m}{B\gamma r}\psi = 0 \tag{8.34}$$

that may be rewritten in terms of the new variable $\bar{r}^2 = r$, i.e.,

$$\frac{d}{d\bar{r}}\left(\frac{1}{\psi}\frac{d\psi}{d\bar{r}}\right) + \frac{4a_m}{B\gamma}\psi = 0. \tag{8.35}$$

We multiply this equation by $(1/\psi)(d\psi/d\bar{r})$ and calculate the first integral of the equation thus obtained. It is given by

$$\left(\frac{1}{\psi}\frac{d\psi}{d\bar{r}}\right)^2 + \frac{4a_m}{B\gamma}\psi = \Delta \tag{8.36}$$

and the exact solution may be written as

$$\psi = \frac{\Delta}{(8a_m/B\gamma)\sinh^2\sqrt{(\Delta r/4)}}. \tag{8.37}$$

Using the above definition, we find that the exact solution for the inhomogeneous distribution of particles is given by

$$\rho(\mathbf{r}) = \frac{\Delta}{8a_m\gamma^2 r^3 \sinh^2\sqrt{(\Delta r/4)}} \tag{8.38}$$

and thus obtain good assumptions concerning the behavior at long distances from the centre of the inhomogeneous particle distribution. This behavior is related to the results obtained earlier in [114–116], where Boltzmann equation was used for the distribution function in the case of a spherical isolated stellar system. The distribution of particles is inhomogeneous for sizes $R = 1/4\Delta$ and divergent toward the centre,

$$\rho(\mathbf{r}) = \frac{1}{2a_m\gamma^2 r^4}.$$

In this case, the energy of the system is conserved. However, we do not know the coefficients. Thus, we propose the approach given below. If particles are concentrated at short distances and their concentration is very high, then quantum effects become a crucial

factor and our approach is inapplicable. This is the degeneration
condition. The relation between the critical temperature and particle
concentration in this quantum case is determined by the natural
condition

$$\Lambda^3(\mathbf{r})\rho(\mathbf{r}) = \left(\frac{\hbar^2 \beta_c}{2me}\right)^{3/2} \rho_c = 1. \tag{8.39}$$

According to this condition the quantum effect constrains the
gravitation collapse in the system and determines the sizes of neutron
stars. This relation along with the formula for the conservation of the
number of particles, $(4\pi)/(2a_m R_c) = N$ determines all the required
parameters, i.e., the critical distance $R_c = (\hbar^2)/(ma_m N^{1/3})$, the
coefficient $\gamma^2 = (2\pi me)/(\hbar^2 N^{2/3})$, the critical temperature $\beta_c = \gamma^2 R_c^2$, and the concentration $\rho_c = (1/(2a_m R_c^4))$. The energy of the
system is in this case given by $E = (3/2)NkT$, i.e., it is equal to
the energy of a free particle. In this section, we present the general
solution for the classical particle distribution at long distances from
the centre of an inhomogeneous cluster of condensed matter that is
subject to the laws of quantum physics. In all cases, our solution
holds under the condition of classical physics, $\Lambda^3(\mathbf{r})\rho(\mathbf{r}) \ll 1$.

   In this section, we describe a system with $\Lambda^3(\mathbf{r})\rho(\mathbf{r}) = \alpha =$
const $\ll 1$. In this case, we can determine only the behavior of
temperature that is governed by the equation

$$\Delta\left(\frac{1}{\sqrt{\beta(\mathbf{r})}}\right) + \frac{a_m e^\alpha}{B \ln \alpha} \frac{1}{\beta} = 0. \tag{8.40}$$

Similarly to the previous case, the solution of this equation may be
written in the form $\beta = \gamma^{-2} r^{-2n}$ and thus we find that it holds for
$n = -2$, i.e., the temperature varies as $kT = \gamma^2 r^{-4}$, the concentration
is varied as $\rho = Ar^{-6}$, and the normalization conditions for the
conservation of particle number and energy are satisfied. The limiting
behavior of the concentration and temperature provides a suitable
solution to the problem in this special case. Finally, we make an
attempt to present an arbitrary solution in the general case of space-
coordinate dependence of the concentration and temperature.

This equation describes any problem associated with inhomogeneous distributions of particles, temperature, and concentration in self-gravitating systems. Indeed, though the equation cannot be solved in the general case, it provides a possibility to analyze many cases of the behavior of a self-gravitating system under various external conditions. We may consider many realistic distributions of concentration, temperature, and field in a gravitation system but should not use the equation of state that exists as a condition related to this situation. In a realistic system, the temperature cannot be related to the particle distribution. It is a thermodynamic parameter that determines a condition for the behavior of the system and can be found from other physical reactions, not only gravitational. In the general case, we cannot obtain the general solution of the present equation, but can believe that this equation governs the thermodynamics of self-gravitating systems.

We may conclude that the new approach in terms of the nonequilibrium statistical operator allowing for inhomogeneous distributions of particles and temperature is efficient. The statistical operator has no singularities for various values of the gravitation field. The approach makes it possible to solve the problem of self-gravitating systems of particles with inhomogeneous distributions of particles and temperature. The equation of state for the self-gravitating system has been determined. A new length of the statistical instability and parameters of the spatially inhomogeneous distribution of particles and temperature are found for realistic gravitational systems. The gravity factor can either promote or retard such transformations depending on the system and the conditions concerned. For the first time, a description is given of the formation of spatially inhomogeneous particle distributions accompanied by the changes of temperature. The statistical description of the system is tailored to treat the gravitating particles from an arbitrary spatial inhomogeneous particle distributions. In this approach, the probable behavior of a self-gravitating system can be predicted for any external conditions. Moreover, the method may also be applied for the further development of physics of self-gravitating and similar systems that are close to equilibrium.

8.2   Systems with Repulsive Interaction

The proposed approach is based on the nonequilibrium statistical operator [96] that is more suitable for the description of interacting systems. Next, a new approach is worked out to describe a system with purely repulsive interaction. The equation of state and all thermodynamic functions are determined by the equations that provide the dominant contribution to the partition function. We have also shown that this approach makes it possible to describe inhomogeneous distributions of particles and to determine the required parameters of the interacting system. The main idea is to provide a detailed treatment of an interacting system by the principles of nonequilibrium statistical mechanics and to show how to introduce the fundamental scalar field into the statistical description of the scattering in the system and thus to provide an illustration of the H-theorem.

We have already mentioned that phenomenological thermodynamics is based on the conservation laws for the average values of physical parameters, i.e., the number of particles, energy, and momentum. Statistical thermodynamics of nonequilibrium systems is also based on the conservation laws, but for the dynamical variables rather than their average values. It represents local conservation laws for the dynamical variables. In order to determine the thermodynamic functions of a nonequilibrium system, a representation of the relevant statistical ensembles is needed allowing for the nonequilibrium states of the system. The concept of Gibbs ensemble may provide a description of nonequilibrium stationary states. In this case, we can determine a nonequilibrium ensemble as a set of systems contained under similar stationary external conditions. To determine a local equilibrium ensemble exactly, we have to find the distribution function or the statistical operator of the system [96]. We assume that nonequilibrium states of the system may be described by the inhomogeneous distribution energy $H(\mathbf{r})$ and the microscopic particle density $n(\mathbf{r}) = \sum_i \delta(\mathbf{r}-\mathbf{r_i})$. Starting from the nonequilibrium statistical operator of the local equilibrium distribution given by (8.1) with the Hamiltonian in the general case (8.3) for the repulsive interaction with $\nu = -1$, the statistical operator [96] may be written in the form

$$Q_l = \int D\Gamma D\psi \times \exp\left\{-\int\left[\beta(\mathbf{r})\frac{p^2(\mathbf{r})}{2m(\mathbf{r})}\right.\right.$$

$$\left.\left. - \eta(\mathbf{r}) - i\sqrt{\beta(\mathbf{r})}\psi(\mathbf{r})\right]n(\mathbf{r})d\mathbf{r}\right\}Q_{\text{int}}, \qquad (8.41)$$

where

$$Q_{\text{int}} = \exp\left\{-\frac{1}{2}\int\left[\beta(\mathbf{r})U(\mathbf{r},\mathbf{r}')\right]^{-1}\psi(\mathbf{r})\psi(\mathbf{r}')d\mathbf{r}d\mathbf{r}'\right\}$$

follows from the interaction written in terms of the field variable. The latter general functional integral can be integrated over the phase space. We substitute the definition expression for the density, then the nonequilibrium statistical operator reduces to

$$Q_l = \int D\psi \int D\Gamma\xi(\mathbf{r}_i)\exp\left\{-\left(\beta(\mathbf{r}_i)\frac{p_i^2}{2m_i} - i\sqrt{\beta(\mathbf{r}_i)}\psi(\mathbf{r}_i)\right)\right\}Q_{\text{int}}, \qquad (8.42)$$

where $\xi(\mathbf{r}) \equiv \exp\eta(\mathbf{r})$ is the chemical activity and should take into account that

$$D\Gamma = \frac{1}{(2\pi\hbar)^3 N!}\prod_i dr_i dp_i.$$

Integration over all the moments reduces the real part of the nonequilibrium statistical operator to a simple expression [3, 8] given by

$$Q_l = \int D\psi Q_{\text{int}}\exp\left\{\int\left[\xi(\mathbf{r})\left(\frac{2\pi m(\mathbf{r})}{\hbar^3\beta(\mathbf{r})}\right)^{3/2}\cos\left(\sqrt{\beta(\mathbf{r})}\psi(\mathbf{r})\right)\right]d\mathbf{r}\right\} \qquad (8.43)$$

For the general case of long-range interaction, such as Coulomb-like or Newtonian gravitation interaction in continuum, the limiting inverse operator should be treated in the operator sense, i.e.,

$$U^{-1}(\mathbf{r},\mathbf{r}') = -L_{\mathbf{r}\mathbf{r}'} = -L_{\mathbf{r}'}\delta(\mathbf{r} - \mathbf{r}').$$

For the case of long-range interaction between particles, the nonequilibrium statistical operator may be written as

$$Q_l = \int D\psi \exp\left\{\int \left[\psi(\mathbf{r})L_{\mathbf{r}\mathbf{r}'}\psi(\mathbf{r}')\right.\right.$$
$$\left.\left. + \xi(\mathbf{r})\Lambda^{-3}(\mathbf{r})\cos\left(\sqrt{\beta(\mathbf{r})}\psi(\mathbf{r})\right)\right]d\mathbf{r}\right\} \qquad (8.44)$$

where

$$\Lambda(\mathbf{r}) = \left(\frac{\hbar^2\beta(\mathbf{r})}{2m(\mathbf{r})}\right)^{1/2}$$

is the thermal de Broglie wavelength at each space point. In the general case, the nonequilibrium statistical operator is given by

$$Q_l = \int D\psi \exp\left\{S(\psi(\mathbf{r}), \xi(\mathbf{r}), \beta(\mathbf{r}))\right\} \qquad (8.45)$$

with the effective nonequilibrium "local entropy" being described by the expression

$$S = \int \left[\psi(\mathbf{r})L_{\mathbf{r}\mathbf{r}'}\psi(\mathbf{r}') + \xi(\mathbf{r})\Lambda^{-3}(\mathbf{r})\cos\left(\sqrt{\beta(\mathbf{r})}\psi(\mathbf{r})\right)\right]d\mathbf{r}. \qquad (8.46)$$

The statistical operator is suitable when applying the efficient methods developed in quantum field theory without additional restrictions on either integration over the field variables or the perturbation theory. The functional $S(\psi(\mathbf{r}), \xi(\mathbf{r}), \beta(\mathbf{r}))$ depends on the field distribution, the chemical activity, and the reciprocal temperature. After that, we can apply the saddle-point method to find the asymptotic value of the statistical operator Q_l as the number of particles N tends to infinity. The dominant contribution is given by the states that satisfy the extremum condition for the function. It is not difficult to show that the saddle-point equation represents the thermodynamic relation and may be reduced to an equation for the field variable, $(\delta S)/(\delta\psi(\mathbf{r})) = 0$, the normalization condition

$$\int \frac{\delta S}{\delta\xi(\mathbf{r})}\xi(\mathbf{r})d\mathbf{r} = N,$$

and the energy conservation law for the system,

$$\int \frac{\delta S}{\delta \beta(\mathbf{r})} \xi(\mathbf{r}) d\mathbf{r} = E.$$

The solution of this equation completely determines all the thermo-dynamic functions and describes the general behavior of interact-ing systems with both spatially homogeneous and inhomogeneous particle distributions. The above set of equations in principle solves the many-particle problem in the thermodynamical limiting case. The spatially inhomogeneous solution of these equations corresponds to the distribution of interacting particles. It is very important to note that only this approach makes it possible to take into account the inhomogeneous distribution of the temperature that depends on the spatial distribution of particles in the system.

From the normalization condition and definition $\int \rho(\mathbf{r}) d\mathbf{r} = N$, we may introduce some new variable that presents the macroscopic density function $\rho(\mathbf{r}) \equiv \Lambda^{-3}(\mathbf{r}) \xi(\mathbf{r}) \cos(\sqrt{\beta(\mathbf{r})}\psi(\mathbf{r}))$. In the case without interaction $\varphi(\mathbf{r}) = 0$ for free particles, and if we write the chemical activity in terms of the chemical potential, $\xi(\mathbf{r}) = \exp(\mu(\mathbf{r})\beta(\mathbf{r}))$, then we obtain from this definition the well-known relation $\beta(\mathbf{r})\mu(\mathbf{r}) = \ln \rho(\mathbf{r})\Lambda^3(\mathbf{r})$ that generalizes the relation of equilibrium statistical mechanics [52, 53]. The equation for energy conservation in this case may be presented in the new form:

$$\frac{1}{2} \int \frac{\rho(\mathbf{r})}{\beta(\mathbf{r})} \left[3 - \sqrt{\beta(r)}\psi(\mathbf{r})\tan\left(\sqrt{\beta(\mathbf{r})}\psi(\mathbf{r})\right) \right] d\mathbf{r} = E. \qquad (8.47)$$

Derivation of the energy-conservation equation over the volume yields a relation for the chemical potential, i.e.,

$$\frac{1}{2} \frac{\rho(\mathbf{r})}{\beta(\mathbf{r})} \left[3 - \sqrt{\beta(r)}\psi(\mathbf{r})\tan\left(\sqrt{\beta(\mathbf{r})}\psi(\mathbf{r})\right) \right] = \frac{\delta E}{\delta V} \frac{\delta V}{\delta N} = \mu(\mathbf{r})\rho(\mathbf{r})$$

$$(8.48)$$

hence the chemical potential is given by

$$\mu(\mathbf{r})\beta(\mathbf{r}) = \frac{3}{2} - \frac{1}{2}\sqrt{\beta(r)}\psi(\mathbf{r})\tan\left(\sqrt{\beta(\mathbf{r})}\psi(\mathbf{r})\right). \qquad (8.49)$$

This approach also provides the equation of state for the system within the context of the thermodynamic relation for pressure $P = (1/\beta)(\delta S/\delta V)$ in the case of energy conservation. The local equation of state is now reduced to

$$P(\mathbf{r})\beta(\mathbf{r}) = \rho(\mathbf{r})\left(\mu(\mathbf{r})\beta(\mathbf{r}) - \frac{1}{2}\right). \qquad (8.50)$$

In the case of an ideal gas, $\psi(\mathbf{r}) = 0$, we have $\mu\beta = 3/2$ and obtain the usual equation of state $P\beta = \rho$ that reproduces the equation of state of the ideal gas. The energy of the system is equal to $E = (3/2)NkT$, this formula is in accordance with the previous well-known results [53]. Within the context of the definition (8.50) we thus conclude that, under the condition $\mu(\mathbf{r})\beta(\mathbf{r}) < 1/2$, there arises negative pressure $P(\mathbf{r}) < 0$ that satisfies the necessary vacuum condition in the cosmology. It holds under the special condition $\sqrt{\beta(r)}\psi(\mathbf{r})\tan\left(\sqrt{\beta(\mathbf{r})}\psi r\right) < 2$ for constant temperature and for the total energy of the system $E < (1/2)NkT$. This condition implies that the energy of each particle is lower than the thermal energy. In this special case, the energy of the system is lower than the total thermal energy of particles, that is impossible. The saddle-point method provides a possibility to select system states whose contributions in the partition function are dominant [1, 3]. The solutions of the equations thus obtained are associated with finite values of the functional and may be regarded as thermodynamically stable particle and field distributions. With these solutions we have to determine the general relation between thermodynamic parameters and their spatial dependence. Thus the spatially inhomogeneous distribution of fields can be unambiguously related to the spatially inhomogeneous particle distribution. In the general approach, all thermodynamic parameters (pressure, chemical potential, density) depend on the space point and are mutually dependent. Actually, this approach extends the mean-field approximation to involve spatially inhomogeneous field distributions. In our case, the field that is introduced describes the nature of the repulsive interaction and can present the fundamental scalar field that corresponds to the scattering in the system.

We introduce a new field variable $\varphi = \sqrt{\beta(r)}\psi(\mathbf{r})$ and take into account the definition of the chemical potential (8.49). Then, we may rewrite the macroscopic density in the form given by

$$\rho(\mathbf{r}) \equiv \Lambda_e^{-3}(\mathbf{r}) \exp\left\{-\frac{1}{2}\varphi(\mathbf{r})\tan\varphi(\mathbf{r})\right\} \cos\varphi(\mathbf{r}), \tag{8.51}$$

where the de Broglie wavelength is renormalized, i.e.,

$$\Lambda_e = \left(\frac{\hbar^2 \beta(\mathbf{r})e}{2m(\mathbf{r})}\right)^{1/2}.$$

After substituting the relation thus obtained in (8.46) we find that the local entropy in terms of the new variable is given by

$$S = \int \left\{ \frac{\varphi(\mathbf{r})}{\sqrt{\beta(r)}} L_{\mathbf{r}\mathbf{r}'} \frac{\varphi(\mathbf{r}')}{\sqrt{\beta(r)}} + \Lambda_e^{-3}(\mathbf{r}) \right.$$

$$\left. \times \exp\left[-\frac{1}{2}\varphi(\mathbf{r})\tan\varphi(\mathbf{r})\right] \cos\varphi(\mathbf{r}) \right\} d\mathbf{r}. \tag{8.52}$$

For constant temperature and equal particle masses the local entropy in the mean-field approximation may be presented as

$$S = \int \left\{ \frac{1}{\beta}\varphi(\mathbf{r}) L_{\mathbf{r}\mathbf{r}'} \varphi(\mathbf{r}') + \Lambda_e^{-3} \right.$$

$$\left. \times \exp\left[-\frac{1}{2}\varphi(\mathbf{r})\tan\varphi(\mathbf{r})\right] \cos\varphi(\mathbf{r}) \right\} d\mathbf{r}. \tag{8.53}$$

Now the equation for the field variable may be rewritten as

$$2L_{\mathbf{r}\mathbf{r}'}\varphi(\mathbf{r}') - \beta\frac{dV(\varphi)}{d\varphi} = 0, \tag{8.54}$$

where the potential energy

$$V(\varphi) \equiv \rho(\mathbf{r}) = \Lambda_e^{-3} \exp\left[-\frac{1}{2}\varphi(\mathbf{r})\tan\varphi(\mathbf{r})\right] \cos\varphi(\mathbf{r}) \tag{8.55}$$

is a function of the field variable in the above form. This potential has a minimum for $3\sin 2\varphi = -2\varphi$. For small values of φ, we have two different solutions, $\varphi = 0$ and $\varphi^2 = 1$. For small φ, the effective potential is given by a very simple expression $V(\varphi) = (1 - \varphi^2)$ and the equation for the field variable reduces to $2L_{\mathbf{rr'}}\varphi(\mathbf{r'}) + 2\beta\varphi(\mathbf{r'}) = 0$. In the general case, the potential energy of the field possesses oscillatory character with decreasing amplitude. This representation with the necessary condition of the cosmological model of natural inflation was proposed in [238]. After that, we can analyze the probable spatial solution for the field variable and the behavior of the field in time. In order to provide this, the knowledge of particle interaction energy with relevant specifics is required. It is well-known from cosmology reasoning that galaxy scattering is associated with the fundamental scalar field. We have shown above that the repulsive statistical interaction can be described by the field $\varphi = \sqrt{\beta(r)}\psi(\mathbf{r})$. We suppose that the introduced fields are fundamental scalar fields responsible for the statistical motivation of the scattering of matter. From this assumption, we find the energy of interaction between two masses located at different space points. As follows from cosmology, two masses scatter with the velocity $v = H(\mathbf{r} - \mathbf{r'})$, where H is the Hubble constant and $\mathbf{r} - \mathbf{r'}$ is the distance between them. The kinetic energy of the relative motion for each mass is given by

$$T = \frac{m(\mathbf{r})}{2} H^2 (\mathbf{r} - \mathbf{r'})^2$$

and thus the energy of interaction between two masses located at different space points is

$$W(\mathbf{r} - \mathbf{r'}) = \frac{m(\mathbf{r})}{2} H^2 (\mathbf{r} - \mathbf{r'})^2 + \frac{m(\mathbf{r'})}{2} H^2 (\mathbf{r} - \mathbf{r'})^2.$$

For homogeneous distribution of masses, the latter expression may be rewritten as $W(\mathbf{r} - \mathbf{r'}) = m(\mathbf{r}) H^2 (\mathbf{r} - \mathbf{r'})^2$. In terms of such interaction energy, the inverse operator is given by

$$L_{\mathbf{rr'}} = \frac{1}{mH^2} \frac{d^2}{dr^2}. \tag{8.56}$$

Having determined the inverse operator, we can present the spatial dependence of the fundamental scalar field as the solution of the equation

$$\frac{2}{mH^2}\frac{d^2\varphi}{dr^2} - \beta\frac{dV(\varphi)}{d\varphi} = 0. \tag{8.57}$$

For small values of φ, the latter transforms to the equation

$$\frac{d^2\varphi}{dr^2} + \beta m H^2\varphi = 0, \tag{8.58}$$

that has a periodic solution $\varphi = \cos(\sqrt{m\beta}Hr)$ with the spatial period $(1/\sqrt{m\beta}H)$. For distances shorter than this value, we may regard the fundamental scalar field to be invariable. However, the fundamental scalar field can change in time. To describe the evolution of such fields, we should present a dynamical equation.

In our case, however, this field induces the repulsive interaction in the system and causes entropy increase. The behavior of the solution of the equation is similar to the behavior that follows from the usual equation for scalar field and the formation of a new-phase bubble of the fundamental scalar field. The present solution can describe the formation of a bubble of a new phase in the theory of inflation of the Universe [24, 78], and the field variable introduced plays the role of the fundamental scalar field and takes into account the repulsive interaction in the system under consideration. In the general presentation, formula (8.46) can describe the condition of new phase formation, the size of the bubble, and other parameters of the thermodynamic behavior of such systems. This nonequilibrium statistical describes only probable dilute structures of such systems, it does not describe metastable states and tells nothing about the time scales in the dynamical theory. In this way, however, we can solve complicated problems of the statistical description of interacting systems. For this purpose, we have to derive a dynamical equation for the field. In this sense, we can use the Ginsburg–Landau equation for the fundamental scalar field in the standard form given by

$$\frac{\partial\varphi(\mathbf{r}, t)}{\partial t} = -\gamma\frac{\delta S}{\delta\varphi(\mathbf{r})}, \tag{8.59}$$

where γ is the dynamical viscosity coefficient [65]. In this case, all the necessary conditions satisfy the thermodynamic relation. We may suppose that the motivation of the Universe dynamics is associated with the entropy increase. The evolution in the nonequilibrium state is governed by the local entropy landscape and the morphological instabilities of the parameter. The dynamics of the system is dissipative, it should result in the decrease of the local entropy. This solution of the obtained equation provides the answer to the question, what is the motive of the scattering of matter. The dynamics of the formed Universe can be only dissipative and in order to describe the existence of the Universe we have to take into account its nonequilibrium conditions.

Interacting particle systems are nonequilibrium *a priori*. Before relaxing into thermodynamic equilibrium, isolated systems with long-range interactions are trapped in nonequilibrium quasi-stationary states whose lifetimes diverge as the number of particles increases. A theory that makes it possible to quantitatively predict the instability threshold for spontaneous symmetry breaking in a class of d-dimensional systems was proposed in [94]. Nonequilibrium stationary states were described in [95]. The authors have drawn a conclusion that three-dimensional systems do not evolve to thermodynamic equilibrium but are trapped in nonequilibrium quasi-stationary states. We propose an approach that provides a possibility to quantitatively predict the particle distribution in a system with special repulsive interaction. In this way, we can solve the complicated problem of the statistical description of systems with special repulsive interactions and introduce a new field variable that reduces this task to the solution of the cosmological problem. Moreover, this method may also be applied for the further development of physics of self-gravitating and similar systems that are not far from equilibrium.

8.3   Saddle States of Nonequilibrium Systems

In all cases, a macroscopic system that interacts with the environment turns into the equilibrium state after the relaxation time.

The properties of such systems are determined by the specifics of each system and characteristics of the environment. The equilibrium state of a separate macrosystem may be determined under ideal conditions [53, 65, 117]. Due to the influence of the environment, the thermodynamic parameters of a separate macrosystem become equal to those of the thermal bath.

Within the context of the fundamental principles of thermodynamics, any macroscopic system embedded in a thermal bath approaches equilibrium during some relaxation time. In the equilibrium state, the properties of the system do not depend on how the equilibrium has been established. The equilibrium state is, however, realized only under certain idealized conditions, so in reality, the properties of the system in a quasi-stationary (steady) state may depend both on the specifics of the interaction of the system with the thermal bath and the characteristics of the bath [53–65]. The same concerns nonequilibrium systems.

In the case of a nonequilibrium system the combined parameter cannot be determined. Nevertheless, such systems may possess equilibrium behavior. As is shown below, it is possible to determine the equilibrium state of the system as a stationary state for which the energy interchange between the separate system and the environment is balanced. The state of the system is the result of the balance between the direct influence of the environment on the system and the degradation process caused by the interaction with the environment. It is not difficult to imagine a macroscopic system that can receive energy from the environment but cannot return it back. This depends on the specifics of the system. Examples of such systems are thermal electrons in semiconductors [96]; a system of photons that diffract from inhomogeneities with the diffraction coefficient depending on the frequency of these photons [239, 240]; a system of high-energy particles that may be born in the course of particle collisions in an accelerator; a system of ordinary Brownian particles, when the friction coefficient depends on the velocity. All these systems exist far from thermal equilibrium and the new state is completely determined by the processes of energy exchange. These systems may be described by the distribution functions of the states that may differ from the well-known distribution functions of thermal states.

There are no well-defined methods for determining nonequilibrium distribution functions that would take into account probable states of the macroscopic system [96]. The standard method makes it possible to describe a nonequilibrium state basing on the information on the equilibrium thermal state and small deviations from this state. The nonequilibrium is in this approach manifest as slight modifications of the equilibrium distribution function. An open system can exist very far from equilibrium but nevertheless manifest stationary behavior. In this section, we consider a certain problem related to the description of a nonequilibrium system and analyze a probable definition of a new stationary state taking into account energy dissipation or absorption along with the degradation processes caused by the interaction with the environment.

It is well-known that any state of a system may be described in terms of distribution functions that determine all thermodynamic properties of macroscopic systems [53, 117]. Actually, the statistical description of a macroscopic system requires the knowledge of only few macroscopic parameters, e.g., energy. Hence, it is a fundamental task to work out a method for the study of the general properties of steady states of open systems and to reveal the conditions for such states to exist. One of the probable ways to solve this general problem may be based on the Gibbs approach [241]. The main purpose is to propose a simple way of describing nonequilibrium systems in the energy space [158] and to formulate a new concept of the solution of the cosmological problem. We start from the formulation of the statistical approach.

The canonical partition function in phase space in the equilibrium case may be written as

$$\rho(q, p)d\Gamma = \exp\left\{\frac{F - H(q, p)}{\Theta}\right\} d\Gamma, \qquad (8.60)$$

where $H(q, p) = \varepsilon$ is the Hamiltonian on the hypersurface of constant energy ε, $d\Gamma = \prod_i dq_i dp_i$ is an element of the phase space, $\Theta = kT$, T is the temperature, and F is the free energy that may be found from the normalization condition

$$\int \exp\left\{\frac{F - H(q,p)}{\Theta}\right\} d\Gamma = 1.$$

As it is shown in [241], the phase space depends only on energy and external parameters. We introduce an additional function $\Sigma = \ln(d\Gamma)/(dE)$, then the canonical partition function reduces to the form given by

$$\rho(\varepsilon)d\varepsilon = C \exp\left\{\frac{F - \varepsilon}{\Theta} + \Sigma(\varepsilon)\right\} d\varepsilon. \tag{8.61}$$

The latter presentation makes it possible to describe the dependence of the distribution function on the energy of the macroscopic system [241]. The normalization condition in this presentation may be written in the form

$$\int c \exp\left\{\frac{F - \varepsilon}{\Theta} + \Sigma(\varepsilon)\right\} d\varepsilon = 1 \tag{8.62}$$

from which one can find the normalization constant allowing the determinant transformation between the phase space and the energy variable. In order to select the states that most contribute to the partition function we employ the condition $(d\Sigma)/(d\varepsilon) = 1/\Theta$ that determines the temperature of the system provided the change of phase space as a function of the system energy is known. Using this definition and taking into account the basic principles of statistical mechanics [65], we come to the conclusion that $\Sigma = \ln(d\Gamma)/(d\varepsilon) = S$ is equal to the entropy of the system. Now, we have to make a very important note: the temperature describes the dependence of the entropy only on energy, but not on any other thermodynamic quantities. We can define the temperature for other situations, but this definition makes no sense without changing the entropy. Another important conclusion is that we can calculate the partition function by integrating over energy. Such integration in this sense implies a continual integral over the energy variable. The extremum of the partition function is realized under the condition $F = E - \theta S$ and any probable deviation from this condition makes a very small contribution to the macroscopic characteristics similarly to the quantum contribution to classical trajectories. The novelty of this presentation consists in the possibility to consider the nonequilibrium systems as Brownian systems in the energy space [158]. Starting from the basic kinetic equations for the distribution function of the macroscopic system in the energy space, we can obtain steady states and fluctuation dissipation relations for the nonequilibrium systems [158].

The energy, as a control parameter of the nonequilibrium system, can be a "slow" parameter that determines the state of the system. In the absence of any other knowledge about the nonequilibrium system, there is no reason to prefer any state of the system determined through energy. The system energy variation determines the state of the system. The nonequilibrium distribution function, similarly to the equilibrium case, may be defined as $\rho(\varepsilon, t)$, which includes the dependence on the energy of the system ε and time. The energy distribution function, in the general case, may be obtained from the basic kinetic equation that presents the evolution of the system during a long period of time and takes into account probable fast processes that might occur in it. In terms of the energy, the basic kinetic equation for the nonequilibrium distribution function may be written as

$$\frac{\partial \rho(\varepsilon, t)}{\partial t} = \int \left\{ W(\varepsilon, \varepsilon') \rho(\varepsilon', t) - W(\varepsilon', \varepsilon) \rho(\varepsilon, t) \right\} d\varepsilon', \qquad (8.63)$$

where $W(\varepsilon, \varepsilon')$ is the probability of a transition between different energies of the system. This basic kinetic equation represents the balance equation for the probabilities of states. The energy presentation of a nonequilibrium process is valid only in the case when this variable is canonical and it is possible to carry out averaging. All solutions of the basic kinetic equation for $t \to \infty$ possess a fundamental property — they reduce to the stationary solution that may be interpreted as an "equilibrium" solution for this system. This stationary solution complies with the H-theorem [242]. In the special case, the basic kinetic equation reduces to the Fokker–Planck equation

$$\frac{\partial \rho(E, t)}{\partial t} = \frac{\partial}{\partial \varepsilon} A(\varepsilon) \rho(\varepsilon, t) + \frac{1}{2} \frac{\partial^2}{\partial \varepsilon^2} D(\varepsilon) \rho(\varepsilon, t) \qquad (8.64)$$

that may be derived from the basic kinetic equation in the approximation of fast changes of the energy growth rate and slow changes of the distribution function as a function of the control parameter. The coefficients $A(\varepsilon)$ and $D(\varepsilon)$ depend on energy and their physical meaning should be specified for various situations. The physical meaning of the coefficients may be clarified when we come back to the dynamic equation for the energy. In the general case, one may suppose that the dissipation equation in the standard form is given by

$$\frac{d\varepsilon}{dt} = f(\varepsilon) + g(\varepsilon)L(t). \tag{8.65}$$

This dissipation equation depends on the external influence and initial conditions. The external influence, first of all, is manifest in the change of the energy of the system that dissipates or is absorbed due to the external action. This process is taken into account by the first part of the present equation that describes the direct influence of the environment on a separate macroscopic system. This part can be obtained from the dynamics of any macroscopic system provided the direct interaction of this system with the environment is completely determined. But this is true not for all cases. In the general case, however, the system exists in a contact with the nonlinear environment. The system energy is changed due to the random influence of the environment and this phenomenon is taken into account by the second part of the equation. The random migration of the system is the result of interaction between this system and all other macroscopic systems, the influence of which randomly changes the energy of the system. The influence of this interaction is not correlated and the correlation between two values of fluctuations at two different moments $\langle L(t)L(t')\rangle = \phi(t - t')$ can be nonzero only for the time interval that is equal to the time of action. The symbol $\langle \ldots \rangle$ implies the statistical averaging of the relevant value. The function $\phi\delta(t - t')$ must have a drastic peak in the vicinity of zero and satisfy the condition $\int \phi(\tau)d\tau = \sigma^2$ for the white noise [242]. The system that cannot come to equilibrium after fast changes of the environment should relax to a new state. This process suggests probable degradation of the system in contact with the environment. The energy presentation of the general dissipation equation takes place also in the case of ordinary Brownian particles. The dynamics of Brownian particles may be described in terms of the velocity v by the Langevin equation

$$\frac{dv}{dt} = -\gamma v + F(t),$$
(8.66)

where γ is the friction coefficient and $F(t)$ is the random force associated with the action of the fluid on a particle under the condition that the average over the equilibrium ensemble vanishes, $\langle F(t) \rangle = 0$, and $\langle F(t)F(t') \rangle = \phi^2 \delta(t-t')$ which satisfies the condition for white noise and describes uncorrelated motion of particles. For a Brownian particle with the energy $\varepsilon = Mv^2/2$ the energy change may be written as

$$\frac{d\varepsilon}{dt} = Mv\frac{dv}{dt} = -2\gamma\varepsilon + \sqrt{2M\varepsilon}F(t),$$
(8.67)

that reproduces the previous equation with $f(\varepsilon) = -2\gamma\varepsilon$, $g(\varepsilon) = \sqrt{\varepsilon}$ and $L(t) = \sqrt{2M}F(t)$. Using the solution of the Langevin equation for the velocity, we obtain [242] $\langle v(\infty) \rangle = (\phi^2/2\gamma) = (kT)/M$ and so $\langle \varepsilon \rangle = (kT)/2$, where the temperature of the thermal bath T is introduced. As a solution of the equation in the approximation disregarding the correlation energy, one can also obtain

$$\sqrt{\langle \varepsilon \rangle^2} = \frac{\sigma^2}{4\gamma} \equiv \frac{\phi^2}{4\gamma}2M = kT,$$

that, similarly to the previous result, completely satisfies the equilibrium condition. Different descriptions of the processes occurring in a nonequilibrium system are equivalent, however, the energy presentation is preferable because it facilitates the determination of the condition for the "equilibrium" states of a nonequilibrium system. This approach is valid for various systems for which the direct influence of the interaction with the environment and probable random nonequilibrium fluctuations can be determined. It is preferable because the energy is the slowest variable on which depends the relaxation of the system.

The true nonlinear Langevin equation should have an equivalent equation for the probability distribution function that considers the relevant physical process. Up to now, two different approaches have been proposed. The first one assumes that the coefficient $g(\varepsilon)$ depends on the energy at the starting point, thus the equation for the nonequilibrium distribution function may be obtained in the Ito form. If this coefficient depends on the energy before and after the transition, then the diffusion equation may be written in the Stratonovich form, i.e.,

$$\frac{\partial \rho}{\partial t} = -\frac{\partial}{\partial \varepsilon}\left[f(\varepsilon)\rho\right] + \frac{\sigma^2}{2}\frac{\partial}{\partial \varepsilon}g(\varepsilon)\frac{\partial}{\partial \varepsilon}g(\varepsilon)\rho. \tag{8.68}$$

In what follows, we employ only Stratonovich presentation, because both presentations can be used [76, 242]. Both equations, in different forms, do not make much sense. The physical interpretation of the processes should be studied. In the case under consideration, different states of any system are determined and can be formed by the previous state and probable future states. The present equation for the nonequilibrium distribution function in such a case may be rewritten in more usual form of the local conservation law for the probability, i.e.,

$$\frac{\partial \rho(E,t)}{\partial t} = \frac{\partial J(\rho(\varepsilon,t))}{\partial \varepsilon}. \tag{8.69}$$

The flow of probability may be written as

$$J = -\left[f(\varepsilon) - \frac{\sigma^2}{2}g(\varepsilon)\frac{\partial}{\partial \varepsilon}g(\varepsilon)\right]\rho + \frac{\sigma^2}{2}g^2(\varepsilon)\frac{\partial}{\partial \varepsilon}\rho. \tag{8.70}$$

The comparison of both Fokker–Planck equations in the energy presentation yields

$$A(\varepsilon) = f(\varepsilon) - \frac{\sigma^2}{2}g(\varepsilon)\frac{\partial}{\partial \varepsilon}g(\varepsilon)$$

and the diffusion coefficient

$$D(\varepsilon) = \frac{\sigma^2}{2}g^2(\varepsilon).$$

The stationary solution of the Fokker–Planck equation for $J(\rho(\varepsilon, t)) = 0$, without flow of probability through each energy point, may be given by

$$\rho_s(\varepsilon) = A \exp\left\{\int_{\varepsilon_0}^{\varepsilon} \frac{2f(\varepsilon')d\varepsilon'}{\sigma^2 g^2(\varepsilon')} - \ln\frac{g(\varepsilon)}{g(\varepsilon_0)}\right\}. \tag{8.71}$$

The equilibrium distribution function as a stationary solution in the general nonequilibrium case may be presented as

$$\rho_s(\varepsilon) = A \exp\left\{-U(\varepsilon)\right\}, \tag{8.72}$$

where

$$U(\varepsilon) = \ln\frac{g(\varepsilon)}{g(\varepsilon_0)} - \int_{\varepsilon_0}^{\varepsilon} \frac{2f(\varepsilon')d\varepsilon'}{\sigma^2 g^2(\varepsilon')}. \tag{8.73}$$

This distribution function has the extremum value of energy that may be found as a solution of the equation

$$U'(\tilde{\varepsilon}) = \frac{1}{D(\varepsilon)}\left[D'(\varepsilon) - f(\varepsilon)\right], \tag{8.74}$$

where $'$ stands for the energy derivative. This equation is equivalent to the equation

$$D'(\tilde{\varepsilon}) = f(\tilde{\varepsilon}), \tag{8.75}$$

that determines the relation between the dissipation in the system and the diffusion in the stationary case and thus completely determines the new "equilibrium" state of the system. The stationary nonequilibrium distribution function is given by

$$\rho_s(\varepsilon) = \exp\left\{-U(\tilde{\varepsilon})\right\}\exp\left\{-U''(\tilde{\varepsilon})\varepsilon^2\right\}, \tag{8.76}$$

where

$$-U''(\tilde{\varepsilon}) = \frac{1}{D(\tilde{\varepsilon})}\left(D''(\tilde{\varepsilon}) - f'(\tilde{\varepsilon})\right).$$

This "equilibrium" distribution function is of Gaussian form. When the dissipation $f(\varepsilon)$ is represented by the nonlinear function of the state, and the diffusion coefficient depends on the energy, many interesting situations, including the noise-induced transition into new, more stable, "equilibrium" states, may be obtained. The probable cases that may be described in this approach are given below.

(a) In the case when the diffusion coefficient $g(\varepsilon) = 1$, irrespective of energy, the stationary solution may be written in the form

$$\rho(\varepsilon) = A \exp\left\{ \int_{\varepsilon_0}^{\varepsilon} \frac{f(\varepsilon)}{\sigma^2} d\varepsilon' \right\}, \tag{8.77}$$

where ε_0 is then intrinsic energy of the system. In the case of a conservative system, for $f(\varepsilon) = 0$, the stationary solution transforms into constant. It should be noted that $\varepsilon = \varepsilon_0$ is not only the intrinsic limit but also the stationary point under the absence of the dissipation energy and random diffusion. Moreover, this point is the attractive point and thus the whole stationary distribution function should have an extremum according to the normalization condition for the distribution function [76]. Only in this case, the distribution function can reproduce the microcanonical distribution function.

(b) Furthermore, in the case when only random diffusion of energy occurs, for $g(\varepsilon) = 1$, the equation for the nonequilibrium distribution function takes the form of the simple diffusion equation, the solution is given by

$$\rho(\varepsilon) = A \frac{1}{\sqrt{4\pi\sigma^2 t}} \exp\left\{ -\frac{(\varepsilon - \varepsilon_0)^2}{4\sigma^2 t} \right\}$$

and describes the migration of the system over the energy. The measure of blueness increases in time according to the law $\langle (\varepsilon - \varepsilon_0)^2 \rangle = 2\sigma^2 t$. This solution represents the evolution of the system that in the initial state is described by the equilibrium distribution function $\rho(\varepsilon) = \delta(\varepsilon - \varepsilon_0)$. All states of the system at the initial time are on the surface of constant energy. The fluctuations of the external medium are manifest in the absence of the microcanonical distribution and the occurrence of the uniform distribution over any energy.

(c) If the diffusion coefficient depends on energy and the energy conservation law holds, $f(\varepsilon) = 0$, then the stationary solution is

given by

$$\rho_s(\varepsilon) = A \exp\left\{-\ln\frac{g(\varepsilon)}{g(\varepsilon_0)}\right\} \tag{8.78}$$

that corresponds to the canonical equilibrium distribution function

$$\rho(\varepsilon) = A \exp\left\{-\beta(\varepsilon - \varepsilon_0)\right\} \tag{8.79}$$

only if $g(\varepsilon) = e^{\beta \varepsilon}$, where β may be presented as the reciprocal temperature of the environment. The latter relation holds only if the diffusion coefficient takes the form

$$D(\varepsilon) = \frac{\sigma^2}{2}g^2(\varepsilon) = \frac{\sigma^2}{2}e^{2\beta\varepsilon}.$$

The physical conditions and specifics of the interaction of the system with the external medium should allow for the degradation processes in the system.

(d) When the presentation of the dissipation energy in the system is $f(\varepsilon) \neq 0$ with regard to the previous presentation of the diffusion coefficient, the equilibrium distribution function may be written in the form

$$\rho_s(\varepsilon) = A \exp\left\{-2\beta\left[2\beta - \frac{f'(\tilde{\varepsilon})}{f(\tilde{\varepsilon})}\right]\varepsilon^2\right\}. \tag{8.80}$$

The diffusion coefficient is a universal characteristic of the environment. The stationary solution for the distribution function can be realized only for special relations between the induced and degradation energies. If the system gets as much energy as possible through the contact with the environment, then the stationary state will not be observed.

The energy presentation may be more illustrative for the ordinary description of equilibrium states. For example, for an ordinary Brownian particle, the stationary solution may be written in the form

$$\rho_s(\varepsilon) = A \exp\left\{-\frac{4\gamma}{\sigma^2}\varepsilon - \ln\sqrt{\varepsilon}\right\} \equiv A\frac{1}{\sqrt{\varepsilon}}\exp(-\beta\varepsilon), \tag{8.81}$$

where we have applied the well-known relation $(2\gamma)/(\sigma^2) = \beta$. Taking into account the normalization condition $\int \rho_s(\varepsilon)d\varepsilon \equiv \int \rho_s(p)dp$, we obtain the equilibrium distribution function in the momentum space as given by

$$\rho_s(p) = A\exp\left\{-\beta\frac{p^2}{2M}\right\} = A\exp\left\{-\frac{Mv^2}{2kT}\right\}. \tag{8.82}$$

The stationary solution completely represents the well-known equilibrium distribution function for ordinary Brownian particles.

The energy dependence of the diffusion coefficient may also be obtained from the linear Langevin equation by using the theory of Markovian processes and taking into account the nonequilibrium fluctuations of various coefficients in the function $f(\varepsilon)$. If the dissipation function may be presented in the form $f(\varepsilon) = \alpha_t e^{\beta\varepsilon}$, then the dissipation Langevin equation may be rewritten in another form, i.e.,

$$\frac{de^{-\beta\varepsilon(t)}}{dt} = -\beta\alpha_t, \tag{8.83}$$

where $\alpha_t = \alpha + \xi_t$ contains a constant part and the part ξ_t that describes the influence of the environmental white noise [76]. The Fokker–Planck equation for the nonequilibrium distribution function may be presented in terms of the new variable $z = e^{-\beta\varepsilon}$ in a simple form given by

$$\frac{\partial\rho(z,t)}{\partial t} = \frac{\partial}{\partial z}[\alpha\beta\rho(z,t)] + \frac{\sigma^2\beta^2}{2}\frac{\partial^2}{\partial z^2}\rho(z,t). \tag{8.84}$$

The stationary solution

$$\rho_s(z) = \exp\left\{\frac{2\alpha}{\sigma^2\beta}z\right\}$$

in the case $\beta\varepsilon > 1$ may be presented as $\rho_s(\varepsilon) = \exp\{-\beta\varepsilon\}$ for $(2\alpha)/(\sigma^2\beta) = 1$. This solution is equal to the solution obtained previously by the standard approach.

In this approach, we can also consider a very simple picture of the motion of Browning particles in a heterogeneous medium. For this matter the characteristic of the medium may be taken into account by different values of the friction coefficient. For example, the motion of a large particle in a suspension, a dust particle in a nonhomogeneous medium, and some other interesting systems in the cases when it is necessary to determine the kinetic properties of nonlinear particle-velocity dependence. Therefore, in the above Langevin equation we have to take into account that $\gamma_t = \gamma + \xi_t$ consists of a constant part γ that determines the average friction coefficient, and the chaotic part ξ_t that describes the influence of random changes in the friction of matter. In the case of white noise, the Fokker–Planck equation may be written as [76]

$$\frac{\partial \rho(\varepsilon, t)}{\partial t} = \frac{\partial}{\partial \varepsilon}\left[\gamma \varepsilon \rho(\varepsilon, t)\right] + \frac{\sigma^2}{2}\frac{\partial^2}{\partial \varepsilon^2}\varepsilon^2 \rho(v, t). \tag{8.85}$$

The stationary solution of this equation is [76]

$$\rho_s(\varepsilon, t) = N\varepsilon^{-\frac{1}{2}\left(\frac{2\gamma}{\sigma^2}+1\right)}, \tag{8.86}$$

which may be verified by the direct substitution of the solution in the previous equation. This stationary solution differs from the solution in the standard case, when both diffusion and friction coefficients do not depend on energy. A similar result

$$\rho_s(v, t) = Nv^{-\left(\frac{2\gamma}{\sigma^2}+1\right)}$$

may also be obtained in the velocity presentation. It is possible to present the situation when the system energy increases but there exists a mechanism that limits the energy. For a process that may be described in terms of the dissipation function $f(\varepsilon) = \gamma\varepsilon - \varepsilon^2$, the second part just takes into account such limiting. The absorption parameter may be presented as $\gamma_t = \gamma + \xi_t$ where the second part describes the random change of the environmental influence. The Fokker–Planck equation takes the form given by [76]

$$\frac{\partial \rho(\varepsilon, t)}{\partial t} = \frac{\partial}{\partial \varepsilon}\left[(\gamma\varepsilon - \varepsilon^2)\rho(\varepsilon, t)\right] + \frac{\sigma^2}{2}\frac{\partial^2}{\partial \varepsilon^2}\varepsilon^2 \rho(v, t). \tag{8.87}$$

The stationary solution of this equation may be written as

$$\rho_s(\varepsilon, t) = N\varepsilon^{-\left(\frac{2\gamma}{\sigma^2}+1\right)}\exp\left\{-\frac{2}{\sigma^2}\varepsilon\right\} \tag{8.88}$$

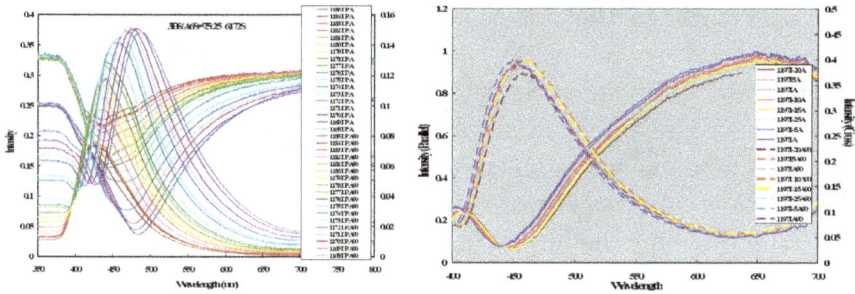

Fig. 8.1. Dependence of the velocity distribution function of charged grains on the charge of moving grain [4, 38, 62].

which resembles though does not reproduce the thermal distribution function. Illustration of this approach describes the behavior of dusty particles in weekly ionized plasma as seen in Fig. 8.1.

A probable approach to the description of nonequilibrium states is proposed. The Fokker–Planck equation for the nonequilibrium distribution function of a macroscopic system is employed to obtain a stationary solution that may be interpreted as the "equilibrium" distribution function for the new energy state. The approach takes into account probable transitions between various states of the system induced by the energy dissipation and influence of the environment that depends on the energy of the system. Nonlinear models are described that represent probable stationary states of a system with various spatial processes in it.

8.4   Nonequilibrium Dynamics of Universe Formation

Standard cosmological models involve a scenario of Universe nucleation and expansion based on the scalar field that is of fundamental importance for the unified theories of weak, strong, and electromagnetic interactions with spontaneous symmetry breaking. An important unexpected feature of the field theory with spontaneous symmetry breaking is that the Universe lifetime in a metastable energy-disadvantageous vacuum state of this field is very long. A theory of new-phase bubble nucleation and expansion was proposed in [243] and extended in [24, 244]. Various cosmological models describe tunneling through the potential barrier in terms of the potentials $V(\varphi)$ of arbitrary forms. At the same time, the cosmological effect of the scalar fields has been proposed as a mechanism to drive the evolution of the Universe in various scenarios. A probable way to solve the fundamental problem in standard cosmology is to consider the inflation Universe of present times. The above-mentioned model takes into account the interaction of a fundamental scalar field with probable fluctuations of fields of other nature. Such fluctuations may be described as the multiplicative vacuum noise. In this case, multiplicative noise not only changes the value of a scalar field, but also modifies the form of the effective potential that modifies the state of the system. This in turn changes the conditions of new phase bubble formation and the evolution of the Universe. The present model differs from the widely studied scenario of the stochastic inflation of the Universe [24, 78] that does not take into account external fluctuations of the fundamental field but allows for the internal field fluctuations of the unstable environment. More realistic models should consider both probable fluctuations as the fundamental scalar field and fields of other nature that arise from the initial condition. The external fluctuation leads to the disorganizing action. The stationary state of the system depends on the adjustments of the potential fluctuations of the environment and the stationary behavior of the system changes [76]. The internal fluctuations change the profile of the potential. However, in order to calculate the size of the bubble we have to violate the equivalence of the local minima. The degree of violation enters the final formula for tunneling probability. The external fluctuations deviate the system from the equilibrium due to intensive interactions of the scalar field with probable fluctuations of the fields of other nature. In this case, the probability to find the system in a given position is presented by the distribution function. The dynamics of the Universe may be described as a stochastic process with a new stationary state formation in the fundamental scalar field that determines the necessary cosmological parameters. In this sense, the new phase can be formed for arbitrary values of the scalar field while the dispersion of fluctuations fully determines the conditions of new phase bubble formation.

A cosmological phenomenon for which fluctuations are crucial is transient dynamics associated with the relaxation from states that have lost their global stability due to the change of appropriate control parameters that in our case is the fundamental scalar field. No unified commonly accepted theoretical approaches have been proposed to determine the probable states of the vacuum. Such systems, however, are abundant in nature and it is a fundamentally important task to develop a method to explore the general properties of the stationary states of the vacuum and to establish the conditions of their existence. A simple model of noise-induced formation of the Universe and its dynamics that allows for the inflation at the present time, is being proposed here.

For the sake of definiteness, we consider the decay of vacuum with different stationary states by the theory of fundamental scalar field that employs the Lagrangian given by

$$L = \frac{1}{2} \left(\partial_\mu \varphi\right)^2 - V\left(\varphi\right), \qquad (8.89)$$

where model potential can be presented in the well-known form:

$$V\left(\varphi\right) = \varepsilon - \frac{\mu^2}{2}\varphi^2 + \frac{\lambda}{2}\varphi^4, \qquad (8.90)$$

where μ^2 is the mass coefficient, λ is the coupling constant. The minimum of the potential in the regime of spontaneous symmetry breaking is characterized by

$$\varphi_0^2 = \frac{\mu^2}{\lambda}, \quad V(\varphi_0) = \varepsilon - \frac{\mu^4}{4\lambda}.$$

In particular, the static Lorentz-invariant Universe requires that the parameter of the potential should be related by $\varepsilon = (\mu^4)/(4\lambda)$. In the limiting case that ignores the matter and radiation but maintains the homogeneity and isotropy of the Universe, the dynamics of the system is governed by the equation [78, 245]

$$3H\dot{\varphi} = -\frac{dV\left(\varphi\right)}{d\varphi}, \qquad (8.91)$$

where $\dot{\varphi}$ denotes time variation and

$$H^2 = \frac{8\pi G V(\varphi)}{3} \qquad (8.92)$$

determines the Hubble parameter that for $\varphi = 0$ is constant, $H = H_0 = \sqrt{(8\pi G \varepsilon)/3}$. If we introduce a new time variable $\tau \equiv t/3H_0$, then the previous dynamic equation for the constant Hubble parameter may be rewritten as

$$\dot{\varphi} = \mu^2 \varphi - \lambda \varphi^3. \qquad (8.93)$$

Now, we suppose that the fundamental scalar field does not completely determine the state of this nonequilibrium system. Such system fluctuations of any other field existing can play an important role in the formation of stationary state. Particularly, fundamental scalar field can interact with such fluctuations that leads to the changes of the state. Moreover, it is shown in [76] that the noise induces a phase transition into the new state and determines the lifetime in the metastable state [246]. The time evolution of the uniform scalar field with regard to the noise is described by the stochastic equation

$$\frac{d\varphi}{dt} = \mu^2 \varphi - \lambda \varphi^3 + \sigma \xi. \qquad (8.94)$$

In this case, the relaxation dynamics of the scalar field depends on the parameter $k = \mu^2/\sqrt{\lambda \sigma}$ that is included in the expressions for the lifetime of the state $\varphi = 0$ and the time dependence of the distribution function $P(x, t)$ [246].

For small noise intensity σ^2, the mean lifetime in the metastable state $\varphi = 0$ may be written as [246]

$$\langle T(0 \to \varphi_0) \rangle = \frac{1}{\sqrt{\lambda \sigma}} \Phi(k) + \frac{1}{2\mu^2} \ln \left| 1 + \frac{\mu^2}{4\lambda \varphi_0} \right|, \qquad (8.95)$$

where $\Phi(k)$ and may be presented by the zero-order modified Bessel function, i.e.,

$$\Phi(k) = \frac{1}{2} \int_0^\infty e^{-ku} K_0(u^2) du = \sum_{n=0}^\infty (-1)^n B_n \frac{k^n}{n!}, \qquad (8.96)$$

where

$$B_n = \frac{\sqrt{2}}{16} 2^{n/2} \left[\Gamma\left(\frac{n+1}{4}\right) \right]^2.$$

If $k \gg 1$ and the barrier height $\Delta V = \mu^4/4\lambda$, then we have $\Phi(k) \approx (1/2k)\ln k^2$ and the mean lifetime yields the value $\langle T \rangle = (1/2\mu^2)\ln(a\varphi_0/\sigma)$. This description is related to the scenario of the interaction of the fundamental scalar field and white noise fluctuations. In nonequilibrium physics, among the phenomena where the effect of statistical fluctuations is crucial, is transient dynamics associated with the relaxation of states that have lost their global stability due to the changes in appropriate control parameter which in the proposed stochastic model is the scalar field. In this sense, we may change the state of the system and change all the parameters that determine the nonequilibrium dynamics of the system. Nevertheless, it is possible to define the new stationary states for such nonequilibrium systems, whose distribution function is different from the well-known equilibrium distributions. Generally, we may assume that the equation that describes the changes of the scalar field in such a nonequilibrium system as the Universe with various fluctuations has the form equivalent to the nonlinear Langevin equation, i.e.,

$$\dot{\varphi} = f(\varphi) + g(\varphi)L(t), \tag{8.97}$$

where

$$f(\varphi) = -\frac{dV(\varphi)}{d\varphi} = \mu\varphi - \lambda\varphi^3.$$

The ground state of any system is determined by fluctuations whose mean values can be zero while the correlations are conserved. Moreover, an arbitrary system can be in contact with a nonlinear environment whose behavior is not fully unambiguous. A random migration of the system over various states is the result of both the direct action of the vacuum and random action due to the contact of the system with the nonlinear vacuum where it is located. A random influence of the vacuum can be taken into account only in the form of correlations between fluctuations at different time instants $\langle L(t)L(t')\rangle = \phi(t - t')$ because the mean values of such fluctuations are equal to zero. The mean value $\langle ... \rangle$ of correlations is nonzero only during the time interval of the action. Therefore, the function $\phi(t-t')$ should have a sharp peak as the time interval tends to zero, which corresponds to the condition $\int \phi(\tau)d\tau = \sigma^2$ that is peculiar to white noise [242]. For this reason, it is more expedient to consider the definition of the nonequilibrium distribution functions of states from the nonlinear Fokker–Planck equation. Such approach is suitable for the description of the behavior of a nonequilibrium system where there occurs relevant energy income from outside, energy losses under the direct action of the environment, and energy dissipation under the random influence of the environment. Considering probable processes that may occur in the system itself and on exchange with the environment, we have to employ a more general approach with nonlinear Langevin equation. Though the form of the Langevin equation differs from that of the Fokker–Planck equation, they are mathematically equivalent [242]. The equation for the distribution function provides more information than the dynamic equation and can describe the probable phase transition to the new state. Assuming that the coefficient $g(\varphi))$ depends on the initial time we have to use the Fokker–Planck equation in the Ito form. If this coefficient depends on the control parameter (in our case, the fundamental scalar field) before and after the transition, then we have to apply the Fokker–Planck equation in the Stratonovich form [242], i.e.,

$$\frac{\partial \rho}{\partial t} = -\frac{\partial}{\partial \varphi}\left(f(\varphi)\rho\right) + \frac{\sigma^2}{2}\frac{\partial}{\partial \varphi}g(\varphi)\frac{\partial}{\partial \varphi}g(\varphi)\rho. \qquad (8.98)$$

In what follows, we only use the Stratonovich representation, the more so, as both approaches are directly related [76, 242]. Both equations have no particular physical meaning until the physical process under consideration is specified. The above equation for the nonequilibrium distribution function of the system may be rewritten by conservation law for the distribution function that is

given by

$$\frac{\partial \rho(\varphi, t)}{\partial t} = \frac{\partial J(\rho(\varphi, t))}{\partial \varphi},$$
(8.99)

where the probability flow may be written in the form

$$J = -\left[f(\varphi) - \frac{\sigma^2}{2} g(\varphi) \frac{\partial}{\partial \varphi} g(\varphi) \right] \rho + \frac{\sigma^2}{2} g^2(\varphi) \frac{\partial}{\partial \varphi} \rho.$$
(8.100)

The general stationary solution of the Fokker–Planck equation in the absence of the flow is given by

$$\rho_s(\varphi) = A \exp \left\{ \int_0^{\varphi} \frac{2f(\varphi')d\varphi'}{\sigma^2 g^2(\varphi')} - \ln g(\varphi) \right\}.$$
(8.101)

The equilibrium distribution function given by the stationary solution associated with the nonlinear properties of the environment may be rewritten in the form

$$\rho_s(\varphi) = A \exp \left\{ -U(\varphi) \right\},$$
(8.102)

where we have introduced the potential

$$U(\varphi) = \ln g(\varphi) - \int_0^{\varphi} \frac{2f(\varphi')d\varphi'}{\sigma^2 g^2(\varphi')}.$$
(8.103)

This distribution function has the form that is not limited by the Gaussian distribution. If the dissipation in the system is described by a nonlinear function $f(\varphi)$ of the scalar field, and the diffusion coefficient depends on the scalar field, a number of situations characterized by new equilibrium states of the nonequilibrium system can be realized. One of the probable ways when considering various fluctuations of the vacuum is to write $\mu^2 = \mu^2 + \xi_2$ where the second term is associated with the random influence of the environment on the given coefficient. The dependence of the diffusion coefficient on the fundamental scalar may be obtained from the linear Langevin equation and using the theory of Markov processes making allowance for the nonequilibrium fluctuations of arbitrary coefficients in the function describing the direct action of the environment $f(\varphi)$. In our case, the second term is associated with the nonlinearity in the system. The standard approach may be given, as before, in terms of the Fokker–Planck equation as [242]:

$$\frac{\partial \rho(\varphi, t)}{\partial t} = \frac{\partial}{\partial \varphi}\left(\mu^2 \varphi - \lambda \varphi^3\right)\rho(\varphi, t)) + \frac{\sigma^2}{2}\frac{\partial^2}{\partial \varphi^2}\varphi^2 \rho(v, t). \qquad (8.104)$$

The stationary solution of the equation is given by

$$\rho_s(\varphi) = N\varphi^{(2\mu^2/\sigma^2)-1}\exp\left\{-\frac{2\lambda\varphi^2}{\sigma^2}\right\}, \qquad (8.105)$$

which corresponds to the non-Gaussian distribution over the scalar field. The extremum of the distribution function now depends on the value of noise. If $0 < \mu^2 < \sigma^2/2$, then the stationary distribution function keeps the behavior of the delta function for $\varphi = 0$. When the parameter becomes greater under the dispersion of the multiplicative noise $\mu^2 > \sigma^2/2$, the distribution function attains maximum for $\varphi = \varphi_0 = \sqrt{\mu^2/\lambda}$ and the Universe is associated with the zero scalar parameter rather than nonzero Hubble parameter. The extremum of the distribution function is in agreement in this case with the zero of the new equation $(\mu^2 - (\sigma^2/2))\varphi - \lambda \varphi^3 = 0$, from which the conclusion follows that the model has two other points of transitions to the new states Fig. 8.2.

The distribution function in the case of simple white noise $g(\varphi) = 1$ from Eq. (1.76) may be presented as

$$\rho_s(\varphi, t) = N\exp\left\{-\frac{2V(\varphi)}{2\sigma^2}\right\}$$

that has an extremum only for $\varphi = \varphi_0$. If we take into account probable dependence of the Hubble parameter as a function, $H = a\sqrt{V(\varphi)}$, then we can solve the Fokker–Planck equation and find the stationary distribution function in the form

$$\rho_s(\varphi, t) = N\exp\left\{-\frac{2H^2}{2a^2\sigma^2}\right\}$$

that has a sharp peak for $\varphi = \varphi_0$ too. This result suggests a conclusion that the current inflation can be explained only by the

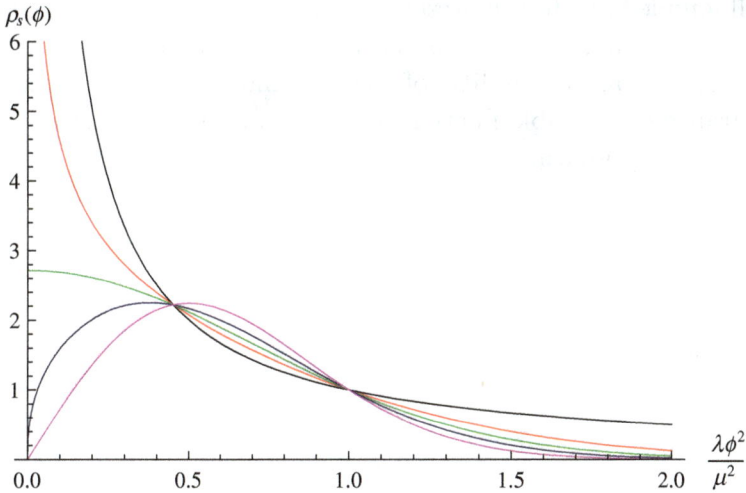

Fig. 8.2.   Dependence of the distribution function on the noise level.

interaction with the multiplicative noise. The state with $\varphi = 0$ can occur here only in the above-mentioned case after observing the necessary condition for the formation of the new phase bubble.

We have shown earlier that when all probable multiplicative fluctuations are taken into account and the probability of a transition into a stable vacuum state is calculated, various situations can be realized. For this reason, we may employ the interaction with the multiplicative noise and regard the probable changes and dispersion of fluctuations as being caused by the scalar field potential. In this case, the nonlinearity should be taken into account both in the potential as in the behavior of fluctuations of other nature. Probable fluctuations change the minimum of the potential and determine an alternative way for dynamical Universe formation. With this behavior, the state of the Universe in the present case can determine $\varphi = 0$ rather than nonzero Hubble constant. Thus, a model is proposed for describing the nonequilibrium Universe by the formation of new stationary states. The stationary distribution functions of the Universe are obtained for the standard mechanism of formation, in contact with the nonlinear environment. This statement is well founded due to the presence of the negative mass coefficient in the fundamental scalar field potential. The negative mass coefficient gives rise to an instability of the ground state and favors the appearance of probable fluctuations that can be taken into account only in this approach.

Conclusions

The results presented in this book illustrate the possibility to describe the systems of interacting particles by their spatially inhomogeneous distributions using statistical theory. The representation of the partition function in terms of the functional integral over the auxiliary fields corresponds to the construction of an equilibrium sequence of probable states with respect to their weights. With the partition function being treated in this way, we may employ the methods of quantum field theory. The extension to the complex plane provides a possibility to apply the saddle-point method and thus to select the states whose contributions in the partition function are dominant. The solutions associated with the finite values of the "effective thermodynamic potential" functional may be regarded as thermodynamically stable particle distributions. Whether the distribution is homogeneous or inhomogeneous depends on the solutions that satisfy the extremum condition for the functional. Thus the spatially inhomogeneous distribution of the auxiliary fields may be unambiguously related to the spatially inhomogeneous particle distribution. It is also possible to find the parameters of such formations and the temperature of the phase transition accompanied by the formation of finite-size clusters of the new phase. Actually, this approach extends the mean-field approximation to consider spatially inhomogeneous field distributions.

The approach proposed here is an advancement in the study of the behavior of clusters formed. In the case of multisoliton solutions, the residual interaction (uncompensated in the course of cluster formation) produces new spatial structures. The soliton interaction energy is described by an expression of the form $w_{rr'} = A \exp[-k(r - r')]$ [33, 65]. Obviously, this system of clusters may be regarded as a gas of interacting particles and traditional methods of statistical physics, e.g., the simplest Ising model, may be employed to estimate the temperature of the phase transition to the spatially ordered state. It is also possible then to consider the collective behavior of such formations.

In the approach considered here, there is no need to introduce two auxiliary fields that correspond to the attraction and repulsion, respectively. We may introduce one complex field $\varphi + i\psi$ associated with the interaction of any type, and carry out the procedure in the complex plane. We only have to know the inverse operator of the interaction. Dividing the interaction into several parts provides a better understanding of the mechanisms of spatially inhomogeneous particle distribution formation. Actually, this method describes the first-order phase transitions to the states that contain new phase bubbles.

In this book, based on the new method, we have studied the properties of the model system of a self-gravitating gas with fixed number of particles N and energy. The size of the cluster is determined by the balance of gravitation and thermal energies, so that with the increase of the temperature, the size of the cluster increases while the density in the center decreases. The size is determined by the minimum of the free energy, however, the relevant minimum is not absolute and the process of collapse may continue further while each cluster acts as a particle. This situation is analogous to the false vacuum in field theory [24]. In the case of large but finite size of the system and the number of particles in it, a value of temperature exists such that if the temperature of the system is higher than this value, then the gas cannot collapse. Moreover, such a value of concentration exists so that if the concentration of the system is lower than this value, then the gas cannot collapse too. The collapse cannot occur for any degree of compression if the temperature is higher than the critical temperature. Our results show that the gravitation interaction of particles leads to the formation of a cluster of finite size, as the initial homogeneous state is unstable in the limiting case $N \to \infty$, $V \to \infty$, but N/V is fixed. The equation for the spatial distribution function is obtained, and the size of the spatial inhomogeneity is determined by the thermodynamic conditions.

The characteristic sizes of the two media, i.e., the gravitating gas and repulsing particles (by Coulomb) are similar to the same interaction constant in the Boltzmann limit and the degenerated case. The reason for this coincidence is that the spatial distributions in both cases are determined by the two energies only, i.e., the Coulomb-type and thermal energy or the repulsive Fermi "statistical potential". The mechanisms of clusters and polarons formation are similar — the balance of the two energies. This fact may be explained in terms of the hydrodynamic approach — both media have equal time scales of the structure formation, but in the first case, it determines the relaxation time while in the second case, it determines the plasma frequency. These time scales determine spatial scales (the radii of clusters and polarons) monotonously.

To solve the problem of the divergence of the partition function for a gravitating system is impossible. We have shown, however, that this difficulty may be avoided and hence the state with multitude of clusters may be treated as metastable. The gravitation interaction of particles results in the cluster formation of finite size, as the initial homogeneous state is unstable. The size is determined by the thermodynamic conditions, in particular, the increase of temperature is accompanied by the increase of the cluster size that causes the decrease of the mean density; the cluster size decreases when particles are added in the system that is associated with the increase of the gravitation energy. Such behavior is caused by the long-range attraction of the gravitation interaction $(1/R)$. The size of the cluster approaches the equilibrium size asymptotically in the course of its formation. The state of a system with a spatially inhomogeneous distribution function corresponding to the cluster of equilibrium size is stable.

Unfortunately, the results of the study of this model cannot be verified experimentally. Nevertheless these results (cluster formations within the Boltzmann limit) may be useful in problems of astrophysics, for the study of the formation of giant planets, stars, and the accumulation of the gas-dust matter, in particular.

We have developed a formalism to describe the spatially inhomogeneous distribution in a system of interacting particles. It employs the new unconventional method that uses the Hubbard–Stratonovich representation of the partition function. The method is extended and applied to a system with Coulomb-like interaction to find the solution for the particle distribution. It is important that this solution has no divergences for the thermodynamic limiting cases. We use the saddle-point approximation by considering the conservation of the number of particles that yields a nonlinear equation for the new field variable. In the three-dimensional case, this equation reduces to the sine-Gordon equation whose solution determines the state associated with the dominant contribution in the partition function. This approach helps to describe the conditions of Wigner crystal formation in a system of dust particles in a plasma. Various possibilities may be probable for different parameters corresponding to the interaction potential. However, the results for simple and basic cases are very important in the understanding of the behavior of a dusty plasma in situations involved. We have analytically derived the necessary condition for the crystal formation in a system of dust particles in the three-dimensional case. In the one- and two-dimensional cases, we have found exact solutions for various spatial distributions of charged particles. The proposed method is designed for the studies of selfassembling systems where nonuniform distributions are found for different scale lengths. We present the partition function and the equation of state in the most general form. In the three-dimensional case we obtained a condition (quadratic dependence of the coupling parameter on the experimental parameter l) that is in good agreement with the experimental data reported in [34]. In the one-dimensional case, we obtained a structure of the periodic distribution of charged particles along a cylindrical sample. In the two-dimensional case, an exact result is obtained for the partition function of the homogeneous distribution in a purely Coulomb system.

The approach proposed in this book makes it possible to describe the first-order phase transitions accompanied by the formation of new-phase macrobubbles at both microscopic and phenomenological levels. The selection of states that bring dominant contributions in establishing the stable thermodynamic behavior, making allowance for fluctuations at both levels, provide a complete picture of processes that accompany the first-order phase transitions. The state selection is combined with the calculation of the saddle-point configuration of the fields that contribute the most towards the thermodynamic characteristics whereas the fluctuations against the background of the order parameter treated as the slowest subsystem enables us to consider fluctuations into account for both cases of small-scale and long-wave variations. The separation of scales occurs under the formation of a finite-size nucleus of the new phase. The fluctuations of the scale smaller than the bubble size may be taken into account by averaging over the fast variable. Thus the treatment is reduced to considering a new effective potential regarding the collective condensate behavior of fast fluctuations. The large-scale fluctuations whose variation scale is larger than the spatial rate of change of the order parameter act as an external field with respect to the bubbles formed, so the dispersion of changes thereof determines the average size of the new phase bubble. Thus it is possible to consider fluctuations of any scale by a unified approach given that fast fluctuations produce the nonlinear potential for the order parameter that just serves to form the new phase bubble, while slower fluctuations determine the size of the latter that can be observed experimentally. Separation of fluctuations of a certain type is governed mainly by the behavioral specifics of the system at the microscopic level; separation of the main governing parameters is associated in the long run with the character and value of the interparticle interaction in the system. The correct treatment of the microscopic behavior of the system determines all the governing parameters of the system along with the characteristics of the first-order phase transition accompanied by the formation of new-phase macrobubbles.

The equilibrium distribution of interacting particles corresponds to their spatially nonuniform distribution. This spatial nonuniform distribution of the introduced particles creates areas free of particles. The combined effect of particles and the medium produces a non-uniform distribution of cooperating particles. Some kind of a new soft body is formed whose properties differ from the properties of the medium. It is possible to form a wide variety of cellular structures if the forces of interaction possess anisotropy. Another important property of such systems is that they are highly sensitive to weak external effects. This enables us to exert a specific effect upon the formation of structures and their transmutations. Moreover, this system has got an advantage of being visually observable because one can directly observe all the changes in the structures. Thus, a general description of the formation of a cellular structure in the system of interacting particles is proposed. The analytic results have been presented for probable cellular structures in ordinary colloidal systems, in systems of particles immersed in a liquid crystal, and in gravitation systems. The formation of cellular structures in all systems of interacting particles for different temperatures and concentrations of particles is shown to have the same physical nature and have the same theoretical description.

Interacting particle systems can be nonequilibrium *a priori*. Before relaxing towards thermodynamic equilibrium, isolated systems with long-range interactions are trapped in nonequilibrium quasi-stationary states whose lifetimes diverge as the number of particles increases. A theory is presented that makes it possible to quantitatively predict the instability threshold for spontaneous symmetry breaking for a class of d-dimensional systems. Nonequilibrium stationary states of three-dimensional systems do not evolve to thermodynamic equilibrium but are trapped in nonequilibrium quasi-stationary states. The presented nonequilibrium statistical description deals only with the probable structures that occur in a self-gravitating system. It does not describe the states and tells nothing about time scales of the kinetic theory. The statistical operator has no singularities for any value of the gravitation field. The challenge to describe a self-gravitating particle system can be solved by this approach provided we take into account inhomogeneous particle and temperature distributions. The gravity factor may either favor or retard such transformations; this depends on the character of the system under consideration and the relevant conditions. For the first time, we have described the formation of spatially inhomogeneous particle distributions followed by the changes of the temperatures of these distributions of interacting particles. The statistical description of the system was tailored to the particles with gravitational interaction by an arbitrary spatially inhomogeneous particle distribution. In this approach, the behavior of a self-gravitating system can be predicted for any external conditions. In this way, one can develop a statistical description of a self-gravitating system. We propose an approach that provides a possibility to quantitatively predict the particle distribution in a system with special repulsive interaction. In this way, we can solve the complicated problem of the statistical description of systems with special repulsive interactions and introduce a new field variable that reduces this task to the solution of the cosmological problem. Moreover, this method may also be applied for the further development of physics of self-gravitating and related systems that are not far from equilibrium.

Another important conclusion is that we can calculate the partition function by integration over the energy. Such integration in this sense implies a continual integral over the energy variable. The extremum of the partition function is realized under the thermodynamic condition and any probable deviation from this condition contributes very little to the macroscopic characteristics similar to the quantum contribution to classical trajectories. The novelty of this presentation consists in the possibility to use the kinetic equations for the description of nonequilibrium systems such as Brownian systems in the energy space. Starting from the basic kinetic equations for the distribution function of the macroscopic system in the energy space, we can obtain steady states and fluctuation dissipation relations for nonequilibrium systems. A probable approach to the description of nonequilibrium states has been proposed too. Based on the Fokker–Planck equation for the nonequilibrium distribution functions of macroscopic systems, a stationary solution was obtained that may be interpreted as the "equilibrium" distribution function for the new energy state. Such approach takes into account probable motion between different states of the system, induced by the dissipation of energy and influence of the environment that depends on the energy of the system. A nonlinear model is proposed representing probable stationary states of systems with various spatial processes within them.

Acknowledgments

This book was prepared with the financial support of the National Academy of Sciences of Ukraine within the research grants "The structure and dynamics of statistical and quantum field systems," "Noise-inducing dynamics and correlations in nonequilibrium systems" and "Mathematical models of non-equilibrium processes in the open systems". The authors are also very grateful to Dr. Olga Kocherga for translating and editing the book into English and to Ms. Zoya Vakhnenko for the help in preparing the manuscript.

Bibliography

[1] E. D. Belotskii, B. I. Lev. *Teor. Mat. Fiz. (Sov)* **60**, 121 (1984).

[2] E. D. Belotskii and B. I. Lev. *Phys. Lett.* **147**, 13 (1990).

[3] B. I. Lev and A. Ya. Zhugaevych. *Phys. Rev. E* **57**, 6460 (1998).

[4] B. I. Lev and A. G. Zagorodny. *Phys. Lett. A* **373**, 1101–1104 (2009).

[5] K. V. Grigorishin and B. I. Lev. *Phys. Rev. E* **71**, 066105 (2005).

[6] B. I. Lev and K. V. Grigorishin. *Cond. Matt. Phys* **45**, 1 (2006).

[7] H. Kleinert. *Gauge Fields in Condensed Matter*. World Scientific, Singapore, 1989.

[8] H. J. de Vega, N. Sanches, and F. Combes. *Phys. Rev. D* **54**, 6008 (1996).

[9] L. N. Lipatov. *JETP* (Sov) **72**, 412 (1977).

[10] S. Edward and A. Lenard, *J. Math. Phys.* **3**, 778 (1962).

[11] J. Hubbard. *Phys. Rev. Lett.* **3**, 77 (1959).

[12] R. L. Stratonovich. *Sov. Phys. Dokl.* **2**, 416 (1984).

[13] S. Samuel. *Phys. Rev. D* **18**, 1916 (1978).

[14] V. B. Magalinsky. *JETP (Sov)* **48**, 167 (1965).

[15] H. Kleinert, Fortschr. *Phys.* **26**, 565 (1979).

[16] K. Rose, E. Gurewitz, and G. C. Fox. *Phys. Rev. Lett.* **65**, 945 (1990).

[17] H. M. Ellerby. *Phys. Rev. E* **49**, 4287 (1994).

[18] K. S. Min, K. V. Shcheglov, C. M. Yang, and H. A. Atwater. *Phys. Rev. Lett.* **69**, 2033 (1996).

[19] B. S. Edelman. *Usp. Fiz. Nauk (Sov)* **130**, 675 (1980).

[20] B. J. Melnikov and C. V. Meschkov, *JETP Lett.* **33**, 222 (1981).

[21] Jo. E. Gegusin, Ju. S. Kaganovskii, and B. B. Kalinin. *Phys. Stat. Solid.* **11**, 250 (1969).

[22] W. Skorupa, R. A. Yankov, and I. E. Tyschenko. *Appl. Phys. Lett.* **68**(17), 2410 (1996).

[23] E. D. Belotskii. Doctor Degree Thesis, *Institute of Physics*, Kiev, 1992.

[24] S. Coleman. *Phys. Rev. D* **15**, 2929 (1977).

[25] S. Ramaswamy, R. Nityananda, V. A. Raghunathan, and J. Prost. *Molecular Crystals and Liquid Crystals* **288**, 175 (1996).

[26] S. Edward and A. Lenard. *J. Math. Phys.* **3**, 778 (1962).

[27] C. C. Bradley, C. A. Sackett, J. J. Tollett, and R. G. Hulet. *Phys. Rev. Lett.* **75**, 1687 (1995).

[28] K. B. Davis, M. O. Mewes, M. R. Andrews, N. J. van Druten, D. S. Durfee, D. M. Kurn, and W. Ketterle. *Phys. Rev. Lett.* **75**, 3969 (1995).

[29] C. C. Bradley, C. A. Sackett, and R. G. Hulet. *Phys. Rev. Lett.* **78**, 985 (1997).

[30] L. P. Pytaevsky. *Zh. Eksp. Teor. Fiz.* **40**, 646 (1961).

[31] E. P. Gross. *J. Math. Phys.* **4**, 195 (1963).

[32] P. A. Ruprecht, M. J. Holland, and K. Burnett. *Phys. Rev. A* **51**, 4704 (1995).

[33] R. Rajaraman. *Phys. Rev.* **15**, 2866 (1977).

[34] V. E. Fortov, A. V. Ivlev, S. A. Khrapak, A. G. Khrapak, and G. E. Morfill. *Phys. Rep.* **421**, 1 (2005).

[35] H. Lowen. *Phys. Rep.* **237**, 249 (1994).

[36] G. E. Morfill, H. M. Thomas, U. Konopka, and M. Zuzic. *Phys. Plasmas* **6**, 1769 (1999).

[37] E. Helpand and F. H. Stilinger. *J. Chem. Phys.* **49**, 1232 (1968).

[38] B. I. Lev, V. B. Tymchyshyn, and A. G. Zagorodny. *Condens. Phys.* **12**, 593 (2009).

[39] H. Thomas, G. E. Morfill, V. Demmel, J. Goree, B. Feuerbacher, and D. Mohlmann. *Phys. Rev. Lett.* **73**, 652 (1994).

[40] J. H. Chu and I. Lin. *Phys. Rev. Lett.* **72**, 4009 (1994).

[41] A. Melzer, T. Trottenberg, and A. Piel. *Phys. Lett. A* **191**, 301 (1994).

[42] A. G. Sitenko, A. G. Zagorodny, and V. N. Tsytovich. *AIP Conf. Proc.* **345**, 311 (1995).

[43] S. A. Brazovsky. *Sov. Phys. JETP* **68**, 715 (1975).

[44] B. I. Lev and H. Yokoyama. *Int. J. Mod. Phys. B* **17**, 4913 (2003).

[45] H. Totsuji, T. Kishimoto, and C. Totsuji. *Phys. Rev. Lett.* **78**, 3113 (1997).

[46] S. V. Vladimirov, S. A. Khrapak, M. Chaudhuri, and G. E. Morfill. *Phys. Rev. Lett.* **100**, 055002 (2008).

[47] H. Ikezi, *Phys. Fluids.* **29**, 1764 (1986).

[48] A. Melzer, A. Homann, and A. Piel. *Phys. Rev. E* **53**, 2757 (1996).

[49] J. C. Crocker and D. G. Grier, *Phys. Rev. Lett.* **77**, 1897 (1996).

[50] R. Aveyard, B. P. Binks, J. H. Clint, P. D. I. Fletcher, T. S. Horozov, B. Neumann, V. N. Paunov, J. Annesley, S. W. Botchway, D. Nees,

A. W. Parker, A. D. Ward, and A. N. Burgess. *Phys. Rev. Lett.* **88**, 246102 (2002).

[51] R. R. Netz. *Phys. Rev. E* **60**, 3174 (1999).

[52] D. Ruelle. *Statistical Mechanics: Rigorous Results.* Benjamin, New York, 1969.

[53] K. Huang. *Statistical Mechanics.* J. Wiley and Sons, New York, 1963.

[54] A. Isihara. *Statistical Mechanics.* State University of New York, 1971.

[55] J. Ortner. *Phys. Rev. E* **59**, 6312 (1999).

[56] A. Ciach, *Phys. Rev. E* **78**, 061505 (2008).

[57] B. I. Halperin and T. C. Lubensky. *Phys. Rev. Lett.* **32**, 292 (1974).

[58] Yu. Kagan, G. V. Shlyapnikov, and J. T. M. Walraven. *Phys. Rev. Lett.* **76**, 2670 (1996).

[59] B. I. Lev and A. G. Zagorodny. *Eur. Phys. J. B* **86**, 422 (2013).

[60] G. B. Whitham. *Linear and Nonlinear Waves.* J. Wiley and Sons, New York, 1974.

[61] G. Stell, K. C. Wu, and B. Larsen. *Phys. Rev. Lett.* **37**, 1369 (1976).

[62] B. I. Lev, V. B. Tymchyshyn, V. K. Ostroukh, and A. G. Zagorodny. *Eur. Phys. J. B* **87**, 253 (2014).

[63] R. Tao, J. T. Woestman, and N. K. Jaggi. *Appl. Phys. Lett.* **55**, 1844 (1989).

[64] R. Tao and J. M. Sun. *Phys. Rev. Lett.* **67**, 398 (1991).

[65] L. D. Landau and E. M. Lifshitz. *Statistical Physics.* Pergamon Press, Oxford, 1959.

[66] J. S. Langer. *Ann. Phys.* **41**, 108 (1967).

[67] D. C. Hobenberg, B. I. Halperin. *Rev. Mod. Phys.* **49**, 435 (1977).

[68] A. G. Khachaturyan. *Theory of Phase Transformation and the Structure of the Solid Solutions.* Nauka, Moscow, 1974 (in Russian).

[69] J. M. Ziman. *Models of Disorder.* Cambridge University Press, Cambridge, 1979.

[70] A. Z. Patashinskii and B. L. Pokrovskii. *Fluctuation Theory of Phase Transitions.* Nauka, Moscow, 1975, 1982 (in Russian). Pergamon Press, Oxford, 1979.

[71] P. M. Platzman and H. Fukuyama. *Phys. Rev. B* **10**, 3150 (1974).

[72] F. M. Peeters and P. M. Platzman. *Phys. Rev. Lett.* **50**, 2021 (1983).

[73] E. M. Lifshitz and L. P. Pitajevskii. *Physical Kinetics.* Nauka, Moscow, 1979 in Russian); Pergamon Press, Oxford, 1981.

[74] A. F. Chaykin and T. C. Lubensky. *Principles of Condensed Matter Physics.* Cambridge University Press, Cambridge, 1995.

[75] A. C. Zefflemoyer (Ed). *Nucleation.* Dekker, New York, 1969.

[76] W. Horsthemke and R. Lefever. *Noise-Induced Transition: Theory, Applications in Physics, Chemistry and Biology*. Springer-Verlag, New York, 1984.

[77] A. H. Guth and E. Weinberg. *Nucl. Phys. B* **212**, 321 (1983).

[78] A. D. Linde. *Particle Physics and Inflationary Cosmology*. Chur: Switzerland, Harwood Academic Publishers, 1990.

[79] G. E. Volovik and U. E. Dzjaloschinskii. *JETP* **75**, 1102 (1978).

[80] P.-H. Chavanis. *Int. J. Mod. Phys. B* **20**, 3113 (2006).

[81] W. C. Saslow. *Gravitational Physics of Stellar and Galactic System*. Cambridge University Press, New York, 1987.

[82] P. Tarazona. *Phys. Rev. A* **31**, 2672 (1985); P. Tarazona. *Phys. Rev. Lett.* **84**, 694 (2000);
P. Tarazona and Y. Rosenfeld. *Phys. Rev. E* **55**, R4873 (1997).

[83] H. Lowen. *Phys. Rev. Lett.* **74**, 1028 (1995).

[84] S. H. Chen and N. M. Amer. *Phys. Rev. Lett.* **51**, 9298 (1983).

[85] B. J. Liang and S. C. Chen. *Phys. Rev. A* **39**, 1441 (1989).

[86] D. A. Soville, W. B. Russel and W. R. Schowaiter. *Colloidal Dispersions*. Cambridge University Press, Cambridge, 1989.

[87] T. Padmanabhan. *Phys. Reports* **188**, 285 (1990).

[88] S. Chandrasekhar. *Principle of Stellar Dynamics*. Dover Publication, New York, 1960.

[89] W. Thirring. *Z. Phys.* **235**, 339 (1970).

[90] C. Sire and P.-H. Chavanis. *Phys. Rev. E* **66**, 046133 (2002).

[91] P.-H. Chavanis. *Phys. Rev. E* **65**, 056123 (2002).

[92] H. J. de Vega and N. Sanchez. *Phys. Lett. B* **490**, 180 (2000);
H. J. de Vega and N. Sanchez. *Nucl. Phys. B* **625**, 409 (2002);
H. J. de Vega and N. Sanchez. *Nucl. Phys. B* **625**, 460 (2002).

[93] V. Laliena. *Nucl. Phys. B* **668**, 403 (2003).

[94] R. Pakter, B. Marcos, and Y. Levin. *Phys. Rev. Lett.* **111**, 230603 (2013).

[95] F. P. C. Benetti, A. C. Ribeiro-Teixeira, R. Pakter, and Y. Levin. *Phys. Rev. Lett.* **113**, 100602 (2014).

[96] D. Zubaryev. *Nonequilibrium Statistical Thermodynamics*. Nauka, Moscow, 1971.

[97] A. R. Denton, N. W. Ashcroft, and W. A. Curtin. *Phys. Rev. E* **51**, 65 (1995).

[98] Y. Rosenfeld, M. Schmidt, H. Lowen, and P. Tarazona. *Phys. Rev. E* **55**, 4245 (1997); Y. Rosenfeld. *Phys. Rev. Lett.* **63**, 980 (1989).

[99] C. Rascon, L. Mederos, and G. Navascues. *Phys. Rev. E* **53**, 5698 (1996).

[100] B. B. Laird and D. M. Kroll. *Phys. Rev. A* **42**, 4810 (1980).

[101] S. C. Kim. *J. Chem. Phys.* **106**, 1146 (1997).

[102] W. K. Keget, H. Retas, and H. N. W. Lekkerkerker. *Phys. Rev. Lett.* **83**, 5298 (1999).

[103] I. Ispolatov and E. G. D. Cohen. *Phys. Rev. Lett.* **87**, 210601 (2001).

[104] B. C. Eu and K. Rah. *Phys. Rev. E* **63**, 031203 (2001).

[105] X. Z. Wang. *Phys. Rev. E* **66**, 031203 (2002).

[106] P. Richard, L. Oger, J. P. Troadee, and A. Gervois. *Phys. Rev E* **60**, 4551 (1999).

[107] C. Caccamo, G. Pellicane, D. Costa, D. Pini, and G. Stell. *Phys. Rev. E* **60**, 5533 (1999).

[108] G. Rybicki. *Astrophys. Space Sci.* **14**, 56 (1971).

[109] K. R. Yawn and B. N. Miller. *Phys. Rev. E* **68**, 056120 (2003).

[110] J. J. Aly. *Phys. Rev. E* **39**, 3771 (1994).

[111] B. I. Lev, S. S. Rozhkov, and A. G. Zagorodny. *EPL* **111**, 26003 (2015).

[112] Y. Levin, R. Pakter, F. B. Rizzato, T. N. Teles, and F. P. C. Benetti. *Phys. Reports* **535**, 1 (2014).

[113] M. Joyce and T. Worrakitpoonpon. *J. Stat. Mech.* **2010**, P10012 (2010).

[114] R. Michie. *Mon. Not. R. Astron. Soc.* **125**, 127M (1963).

[115] R. Michie and P. H. Bodenheimer. *Mon. Not. R. Astron. Soc.* **126**, 269M (1963).

[116] W. Jaffe. *Mon. Not. R. Astron. Soc.* **202**, 995J (1983).

[117] R. Balesku. *Equilibrium and Nonequilibrium Statistical Mechanics.* J. Wiley and Sons, New York, 1978.

[118] J. A. S. Lima, R. Silva, and J. Santos. *Astron. Astrophys.* **396**, 309 (2002).

[119] A. P. Boss. *Science* **276**, 1836 (1997).

[120] S. M. Fall. Annals of the New York Academy of Science **336**, 172 (2006).

[121] B. I. Lev. *Int. J. Mod. Phys. B* **25**, 2237 (2011).

[122] D. C. Wang and A. P. Gast. *Phys. Rev. E* **59**, 3964 (1999); C. Rascon, L. Mederos and G. Navascues. *Phys. Rev. E* **53**, 5698 (1996).

[123] S. Hess and M. Kroger. *Phys. Rev. E* **64**, 011201 (2001).

[124] U. F. Edgal and D. L. Huber. *J. Chem. Phys.* **108**, 1578 (1998).

[125] C. N. Likos, A. Lang, M. Watzlawek, and H. Lowen. *Phys. Rev. E* **63**, 031206 (2001).

[126] N. M. Dixit and C. F. Zukoski. *Phys. Rev. E* **64**, 041604 (2001).

[127] J. Bergenholtz and M. Fuchs. *Phys. Rev. E* **59**, 5706 (1999).

[128] B. Groh. *Phys. Rev. E* **61**, 5218 (2000).
[129] V. Kobelev and B. Kolomeisku. *J. Chem. Phys.* **116**, 7589 (2002).
[130] J. A. Anta and S. Lago. *J. Chem. Phys.* **116**, 10514 (2002).
[131] J. H. Herrera, H. Ruiz-Estrada, and L. Blum. *J. Chem. Phys.* **104**, 6327 (1996).
[132] V. Babin, A. Ciach, and M. Tasinkevych. *J. Chem. Phys.* **114**, 9685 (2001).
[133] J. M. Kosterliz and D. J. Thouless. *J. Phys. C* **6**, 1181 (1973); K. J. Stranburg. *Rev. Mod. Phys.* **60**, 161 (1988).
[134] P. G. de Gennes and J. Prost. *The Physics of Liquid Crystals.* Clarendon Press, Oxford (ed. 2), 1993.
[135] F. Brochard and P. G. De Gennes. *J. Phys. (Paris)* **31**, 691 (1970).
[136] P. Aviles and Y. Giga. *Proc. Center Math. Anal. Austr. Nat. Univer.* **12**, 1 (1987).
[137] P. Aviles and Y. Giga. *Proc. Roy. Soc. Edin. Sect. A* **129**, 1 (1999).
[138] G. Gioia and M. Ortiz. *Adv. Appl. Mech.* **33**, 119 (1997).
[139] M. Ortiz and G. Gioia. *J. Mech. Phys. Solids* **42**, 531 (1994).
[140] N. M. Ercolani, R. Indik, A. C. Newell, and T. Passot. *J. Nonlin. Sci.* **10**, 223 (2000).
[141] R. A. Cowley. *Phys. Rev. B* **13**, 4877 (1985).
[142] M. Cross and A. Newell. *Physica D* **10**, 299 (1984).
[143] A. Newell, T. Passot, C. Bowman, N. Ercolani, and R. Indik. *Physica D* **97**, 185 (1996).
[144] A. E. Lobkovsky and T. A. Witten. *Phys. Rev. E* **55**, 1577 (1997).
[145] S. B. Goryachev. *Phys. Rev. Lett.* **72**, 1815 (1994).
[146] B. H. Liu and L. McLerran. *Phys. Rev. D.* **46**, 2668 (1992).
[147] Y. Rubin and M. Gifferman. *Phys. Rev. A.* **29**, 1496 (1984).
[148] R. Feynman. *Statistical Mechanics.* W.A. Benjamin, California, 1972.
[149] R. J. Baxter. *Exactly Solved Models in Statistical Mechanics.* Academic Press, London, New York, Sydney, Tokyo, Toronto, 1982.
[150] S. F. Edwards. *Phil. Mag.* **4**, 1171 (1959).
[151] Y. S. Wu. *Phys. Rev. Lett.* **73**, 922 (1994).
[152] F. D. M. Haldane. *Phys. Rev. Lett.* **67**, 937 (1991).
[153] V. V. Krasnogolovetz and B. I. Lev. *Ukr. Fiz. J.* **39**, 296 (1994).
[154] B. I. Lev. *Phys. Rev. E* **58**, R2681 (1998).
[155] S. Doniach. *Phys. Rev. B* **24**, 5063 (1981).
[156] P. Kampf and T. Zimanyi. *Phys. Rev. B* **47**, 279 (1993).
[157] J. Jeans. *Astronomy and Cosmogony.* Cambridge Univ. Press, 1928.
[158] B. I. Lev. *Eur. Phys. J. Special Topics* **216**, 37–48 (2013).
[159] P. Arnold. *Phys. Rev. D.* **46**, 2628 (1992).

[160] A. D. Linde. *Pep. Prog. Phys.* **42**, 389 (1979).

[161] A. P. Rebesh and B. I. Lev. *Phys. Lett. A* **218**, 1–6 (2017).

[162] V. I. Klyatskin and D. Gurarie. *Usp. Fiz. Nauk* **169**, 171 (1999).

[163] B. B. Kadomtzev and M. B. Kadomtzev. *Usp. Fiz. Nauk* **167**, 649 (1997).

[164] Yu. Kagan, A. M. Muryshev, and G. V. Shlyapnikov. *Phys. Rev. Lett.* **81**, 933 (1998).

[165] Yu. Kagan, E. L. Surcov, and G. V. Shlyapnikov. *Phys. Rev. A* **54**, R1753 (1996).

[166] A. Gammal, L. Tomo, and T. Frederco. *Phys. Rev. A* **66**, 043619 (2002).

[167] G. S. Jeon, L. Yn, S. W. Rhee, and D. J. Thouless. *Phys. Rev. A* **66**, 011603(R) (2002).

[168] W. Ketterle. *Usp. Fiz. Nauk* **173**, 1339 (2003).

[169] L. Parker and Y. Zhang, *Phys. Rev. D* **44**, 2421 (1991).

[170] V. A. Antonov. *Vestn. Leningr. Gos. Univ.* **7**, 135 (1962).

[171] D. Lynden-Bell and R. Wood. *Mon. Not. R. Astron. Soc.* **138**, 495 (1968).

[172] M. Kiessling. *J. Stat. Phys.* **55**, 203 (1989).

[173] B. Stahl, M. Kiessling, and K. Schindler. *Planet. Space Sci.* **43**, 271 (1995).

[174] E. B. Aronson and C. J. Hansen. *Astrophys. J.* **177**, 145 (1972).

[175] E. Follana and V. Laliena. *Phys. Rev. E* **61**, 6270 (1999).

[176] V. Latora, A. Rapisarda, and S. Ruffo. *Phys. Rev. Lett* **80**, 692 (1998).

[177] M. Antoni and A. Torcini. *Phys. Rev. E* **57**, R6233 (1998).

[178] A. J. Vainschtein, V. I. Zakharov, and M. A. Shifman. *Usp. Fiz. Nauk* **136**, 553 (1982).

[179] J. Ruiz-Garsia, R. Gamez-Corrawe, and B. I. Ivlev. *Phys. Rev. E* **58**, 960 (1998).

[180] V. G. Anderson, E. M. Terentjev, S. P. Meeker, J. Crain, and W. G. K. Poon. *Eur. Phys. E* **4**, 11 (2001).

[181] V. G. Anderson and E. M. Terentjev. *Eur. Phys. E* **4**, 21(2001).

[182] J. Loudet, P. Barois, and P. Poulin. *Nature* **407**, 611 (2000).

[183] P. Poulin, H. Stark, T. C. Lubensky, and D. A. Weitz. *Science* **205**, 5770 (1997).

[184] P. Poulin and D. X. Weitz. *Phys. Rev. E* **57**, 626 (1998).

[185] V. Nazarenko, A. Nych, and B. Lev. *Phys. Rev. Lett* **87**, 075504 (2001).

[186] A. Nych, U. Ognysta, M. Skarabot, M. Ravnik, S. Zumer, and I. Musevic. Nature Communications **4**, 1489 (2013).

[187] R. W. Ruhwandl and E. M. Terentjev. *Phys. Rev. E* **55**, 2958 (1997).

[188] S. B. Chernyshuk, B. I. Lev, and H. Yokoyama. *Zh. Eksp. Teor. Fiz. [Sov. Phys. JETP]* **120**, 431 (2001).

[189] B. I. Lev. *Ukr. J. Phys. Reviews* **2**(1), 3–34 (2005).

[190] B. I. Lev, P. M. Tomchuk, and K. M. Aoki. *Condens. Matter Phys.* **6**, 166 (2005).

[191] B. I. Lev, S. B. Chernyshuk, P. M. Tomchuk, and H. Yokoyama. *Phys. Rev. E* **65**, 021603 (2002).

[192] S. B. Chernyshuk, B. I. Lev, and H. Yokoyama. *Phys. Rev. E* **71**, 062701 (2005).

[193] B. I. Lev, K. M. Aoki, and H. Yokoyama. Molecular Crystals and Liquid Crystals Science and Technology. Section A: Molecular Crystals and Liquid Crystals **367**, 537–544 (2001).

[194] F. Ghezzi and J. C. Earnshaw. *J. Phys.: Condens. Matter* **9**, L517 (1997).

[195] A. Borstnik, H. Stark, and S. Zumer. *Phys. Rev. E* **60**, 4210 (1999).

[196] P. Galatola and J. B. Fournier. *Phys. Rev. Lett.* **86**, 3915 (2001).

[197] J. B. Fournier and P. Galatola. *Phys. Rev. E* **65**, 031601 (2002).

[198] T. C. Lubensky, D. Pettey, N. Currier, and H. Stark. *Phys. Rev. E* **57**, 610 (1998).

[199] D. Pettey, T. C. Lubensky, and D. R. Link. *Liquid Crystals* **25**, 579 (1998).

[200] S. L. Lopatnikov and V. A. Namiot. *Zh. Eksp. Teor. Fiz.* **75**, 361 (1978).

[201] P. Poulin, V. Cabuil, and D. A Weitz. *Phys. Rev. Lett.* **79**, 4862 (1997).

[202] P. Poulin, V. A. Raghunathan, P. Richetti, and D. Roux. *J. Phys. II France*, **4**(9), 1557 (1994).

[203] S. V. Burylov and Yu. L. Raikher. *J. Magn. Magn. Mater.* **122**, 62 (1993).

[204] S. V. Burylov and Yu. L. Raikher. *Phys. Rev. E* **50**, 358 (1994).

[205] I. Janossy, A. D. Lloyd, and B. S. Wherrett. *Mol. Cryst. Liq. Cryst.* **179**, 1 (1990).

[206] D. Hilbert. *Die Grundlagen der Physik.* Nachrichten von der Gesellschaft der Wissenschaften zu Göttingen — Mathematisch-Physikalische Klasse. Nachrichten, 395 p., 1915.

[207] A. Einstein. *Die Grundlage der Allgemeinen Relativitatstheorie.* Annalen der Physik **354**(7) 769–822 (1916).

[208] S. Weinberg. *Rev. Mod. Phys.* **46**, 255 (1974).

[209] M. Ikeda and Y. Miyachi. *Prog. Theor. Phys.* **27**, 474 (1962).

[210] T. T. Wu and C. N. Yang. *Properties of Matter under Unusual Conditions.* (H. Mark and S. Fernbach, Eds.). Wiley-Interscience, New York, 1969.

[211] R. Franzosi, M. Pettini, and L. Spinelli. *Phys. Rev. Lett.* **84**, 2774 (2000).

[212] L. Casetti and A. Macchi. *Phys. Rev. E* **55**, 2539 (1997).

[213] L. Caiani, L. Casetti, C. Clementi, G. Pettini, M. Pettini, and R. Gatto. *Phys. Rev. E* **57**, 3886 (1998).

[214] R. Franzosi, L. Casetti, L. Spinelli, and M. Penttini. *Phys. Rev. E* **60**, R5009 (1999).

[215] L. Casetti, R. Livi, and M. Pettini. *Phys. Rev. Lett.* **74**, 375 (1995).

[216] J. Jurkowski. *Phys. Rev. E* **62**, 1790 (2000).

[217] K. Kaviani, A. Dalafi-Rezaie. *Phys. Rev. E* **60**, 3520 (1999).

[218] G. Ruppeiner. *Phys. Rev. A* **44**, 3583 (1991).

[219] G. Ruppeiner. *Phys. Rev. E* **57**, 5135 (1998).

[220] V. Periwal. *Phys. Rev. Lett.* **78**, 4671 (1997).

[221] R. Blumenfeld. *Phys. Rev. Lett.* **78**, 1203 (1997).

[222] G. Garcia de Polavieja. *Phys. Rev. Lett.* **81**, 1 (1998).

[223] W. T. Gozdz and G. Gompper. *Phys. Rev. E* **59**, 4305 (1999).

[224] G. Foltin. *Phys. Rev. E* **57**, R3742 (1998).

[225] M. V. Jarić (ed.). *Aperiodicity and Order. Volume 1: Introduction to Quasicrystals.* Academic Press, 1988.

[226] M. Seul and D. Adelman. *Science* **267**, 476 (1995).

[227] B. I. Lev. *Mod. Phys. Lett. B* **27**, 1330020 (2013).

[228] E. L. Altschuler, T. J. Williams, E. R. Ratner, F. Dowla, and F. Wooten. *Phys. Rev. Lett.* **72**, 2671 (1994).

[229] E. L. Altschuler, T. J. Williams, E. R. Ratner, R. Tipton, R. Stong, F. Dowla, and F. Wooten. *Phys. Rev. Lett.* **78**, 2681 (1997).

[230] M. Bowick, A. Cacciuto, D. R. Nelson, and A. Travesset. *Phys. Rev. Lett.* **89**, 185502 (2002).

[231] V. S. Édel'man. *Levitated Electrons. Sov. Phys. Usp.* **23**, 227–244 (1980).

[232] H. Tanaka, M. Isobe, and J. Yamamoto. *Phys. Rev. Lett.* **89**, 168303 (2002).

[233] S. Cotsakis, P. Leach, and G. Flessas. *Phys. Rev. D* **49**, 6489 (1994).

[234] F. C. MacKintosh and T. C. Lubensky. *Phys. Rev. Lett.* **67**, 1169 (1991).

[235] J. D. Noh and D. Kim. *Phys. Rev. E* **56**, 355 (1997).

[236] B. I. Lev. *J. Mod. Phys.* **7**, 2366 (2016).

[237] B. I. Lev. *Mod. Phys. Lett. B* **27**, 1330020 (2013).

[238]   F. C. Adams, J. R. Bond, K. Freese, J. A. Frieman, and A. V. Olinto. *Phys. Rev. D* **47**, 426 (1993).

[239]   C. W. Gardiner and P. Zoller. *Quantum Noise*. Springer, 2000.

[240]   D. F. Wells and G. J. Milburn. *Quantum Optics*. New Zealand, 2001; D. F. Wells and G. J. Milburn. *Quantum Optics*. Springer Science & Business Media (2nd ed.), 2008.

[241]   J. W. Gibbs. *Elementary Principles in Statistical Mechanics: Developed with Special Reference to the Rational Foundation of Thermodynamics*. Cambridge University Press, New York, 1902.

[242]   N. G. van Kampen. *Stochastic Process in Physics and Chemistry*. North-Holland, Amsterdam, 1990.

[243]   A. D. Linde. *Phys. Lett. B* **108**, 389 (1982); **114**, 431 (1982); **129**, 117 (1983).

[244]   A. Albrecht and P. J. Steinhard. *Phys. Rev. Lett.* **48**, 1220 (1982).

[245]   J. W. van Holten. *Phys. Rev. Lett.* **89**, 201301 (2002).

[246]   P. Colet, F. De. Pasquele, and M. San Miguel. *Phys. Rev. A* **43**, 5296 (1991).

Index